The Acid Rain Debate

Westview Special Studies

The concept of Westview Special Studies is a response to the continuing crisis in academic and informational publishing. Library budgets are being diverted from the purchase of books and used for data banks, computers, micromedia, and other methods of information retrieval. Interlibrary loan structures further reduce the edition sizes required to satisfy the needs of the scholarly community. Economic pressures on university presses and the few private scholarly publishing companies have greatly limited the capacity of the industry to properly serve the academic and research communities. As a result, many manuscripts dealing with important subjects, often representing the highest level of scholarship, are no longer economically viable publishing projects--or, if accepted for publication, are typically subject to lead times ranging from one to three years.

Westview Special Studies are our practical solution to the problem. As always, the selection criteria include the importance of the subject, the work's contribution to scholarship, its insight, originality of thought, and excellence of exposition. We accept manuscripts in camera-ready form, typed, set, or word processed according to specifications laid out in our comprehensive manual, which contains straightforward instructions and sample pages. The responsibility for editing and proofreading lies with the author or sponsoring institution, but our editorial staff is always available to answer questions and provide guidance.

The result is a book printed on acid-free paper and bound in sturdy, library-quality soft covers. We manufacture these books ourselves using equipment that does not require a lengthy make-ready process and that allows us to publish first editions of 300 to 1000 copies and to reprint even smaller quantities as needed. Thus, we can produce Special Studies quickly and can keep even very specialized books in print as long as there is a demand for them.

About the Book and Editors

This collection of essays by noted academicians, lawyers, energy agency administrators, and research analysts focuses on the political and legal aspects of the acid rain debate, the policy options for resolving the controversy, and the international dimensions of acid rain control. The contributors highlight concerns drawn primarily from the developing study of acid rain in political science, economics, public administration, and policy analysis--concerns that are the focal point of the public debate over the nature, impact, and cost of acid rain and the mitigation of its effects. The book complements the impressive body of research from the natural sciences and responds to the need for applied study to help resolve the current policy stalemate on this critical environmental issue. The Acid Rain Debate features a comprehensive annotated bibliography on acid rain and relevant social science research.

Ernest J. Yanarella, associate professor of political science at the University of Kentucky, is coauthor of Energy and the Social Sciences (Westview, 1982). Randal H. Ihara is a professional staff member with the Senate Democratic Policy Committee.

The Acid Rain Debate
Scientific, Economic, and Political Dimensions

edited by
Ernest J. Yanarella
and Randal H. Ihara

Westview Press / Boulder and London

Westview Special Studies in Science, Technology, and Public Policy

All rights reserved. No part of this publication may be reproduced or transmitted in any form or by any means, electronic or mechanical, including photocopy, recording, or any information storage and retrieval system, without permission in writing from the publisher.

Copyright © 1985 by Westview Press, Inc., except for Chapter 7, which is in the public domain.

Published in 1985 in the United States of America by Westview Press, Inc.; Frederick A. Praeger, Publisher; 5500 Central Avenue, Boulder, Colorado 80301

Library of Congress Cataloging in Publication Data
Main entry under title:
The Acid rain debate.
    (Westview special studies in science, technology, and public policy)
    1. Acid rain--Environmental aspects--United States--Addresses, essays, lectures. 2. Acid rain--Environmental essays, lectures. I. Yanarella, Ernest J. II. Ihara, Randal H.
TD196.A25A2835   1985   363.7'386'0973   85-10676
ISBN: 0-8133-7065-5

Composition for this book was provided by the editors
Printed and bound in the United States of America

10 9 8 7 6 5 4 3 2 1

Dedicated to

Elizabeth, Lisa, John, J.P., and Puff

and

Robin, Chris, Nathan, Justin, Cricket,
Jethro, Bitsy, and Caspar

Contents

List of Tables..xi
List of Figures.....................................xiii
Preface and Acknowledgments..........................xv

INTRODUCTION..1

1. An Overview of the Acid Rain Debate:
   Politics, Science, and the Search
   for Consensus--**Randal H. Ihara**...................3

PART I. THE POLITICAL CONTEXT.......................35

2. The Foundations of Policy Immobilism Over
   Acid Rain Control--**Ernest J. Yanarella**..........39

3. Public Opinion and the Environment: The
   Problem of Acid Rain--**Phillip W. Roeder**
   and **Timothy P. Johnson**..........................57

PART II. POLICY ISSUES AND ALTERNATIVES..............81

4. Perspectives on Acid Deposition Control:
   Science, Economics, and Policymaking--
   James L. Regens and Robert W. Rycroft...........87

5. Acid Rain Legislation and the Clean Air
   Act: Time to Raise the Bridge or Lower the
   River?--**Larry B. Parker and John E. Blodgett**...107

6. Predicting Deposition Reductions Using Long-
   Range Transport Models: Some Policy
   Implications--**Glenn P. Gibian**..................137

7. The Design of Cost-Effective Strategies to
   Control Acidic Deposition--**David G. Streets**....173

8. Equitably Reducing Transboundary Causes of Acid Rain: An Economic Incentive Regulatory Approach--**David J. Webber**......................219

PART III. INTERNATIONAL-COMPARATIVE DIMENSIONS....239

9. Environmental vs. Ecological Perspectives on Acid Rain: The American Environmental Movement and the West German Green Party--**Ernest J. Yanarella**...........................243

10. Acid Rain-Acid Diplomacy--**John E. Carroll**.....261

CONCLUSION: A PROSPECTUS ON THE FUTURE............275

11. The U.S. Politics of Acid Rain--**George Clemon Freeman, Jr**............................277

APPENDIX..315

Acid Rain and the Social Sciences: A Selected Bibliography--**Elizabeth W. Yanarella and Ernest J. Yanarella**............................317

A Note on the Contributors........................333

INDEX...337

Tables

1.1--National U.S. current and projected SO_2 and NO_x emissions............................10
3.1--Surveys reviewed in this chapter..............61
4.1--Percent change in electricity rates for 10 million ton reduction...............97
4.2--Distributional impact of selected options for financing acid deposition reductions.......................................99
5.1--Impact of regulation on SO_2 emissions......................................118
5.2--Impact of alternative generating sources for SO_2..................................120
5.3--Forecast of available electric generating capacity, 1983-2030, probable case..............127
5.4--Assumptions for four scenarios..............128
6.1--Example portion of a transfer matrix.........139
6.2--Calculated deposition rates..................139
6.3--Percent reduction in sulfur dioxide emissions under various control strategies for the United States................141
6.4--Predicted and observed current wet sulfate deposition..............................144
6.5a-d--Percent reduction in sulfur deposition and efficiency of various control strategies.....................149
6.6a-d--pH of rainfall predicted to occur under control strategies..................153
6.7--Percent reduction in sulfur dioxide emissions under various control strategies.......................................158
7.1--Summary of 1980 emissions of the two major acid rain precursors by source category and region............................176
7.2--Comparison of emission reductions and and costs (1980 dollars) to comply with alternative control proposals...................188

xi

7.3--Quantitative findings of the
targeted strategy analysis......................191
7.4--Emissions reductions and increased costs
to base case for our control scenarios..........203
7.5--Summary of potential benefits of
alternative bubble options......................210
8.1--1980 state sulfur dioxide and nitrogen
oxide emissions by state........................222
8.2--Top 50 plants by state........................226
8.3--H.R. 3400 required sulfur dioxide
emissions by state................................228

Figures

1.1--Annual mean value of pH........................8
2.1--Competing political alliances.................42
4.1--Annual mean value of pH in
     precipitation..................................91
4.2--Sulfur dioxide and nitrogen
     oxides emissions...............................93
5.1--Impact of electricity demand
     growth on SO_2 emissions....................118
5.2--Domestic nuclear capacity,
     1982-2020.....................................120
5.3--Estimated emissions as a function
     of retirement age of powerplants..............122
5.4--Coal sulfur delivery trends for
     electric utilities in selected
     regions and in the United States..............123
5.5.--One alternative view of future
     utility emission trends.......................125
5.6--Simulations of future emissions..............130
5.7--Simulations of two retrofit programs
     with two revised NSPS programs................132
7.1--Geographical dimensions of the acid
     deposition issues in the eastern U. S.........174
7.2--Hypothetical control-cost function
     for an emission source........................179
7.3--Methodology for determining the
     least-cost strategy for a group of
     sources.......................................182
7.4--Comparison of the costs of alternative
     control proposals with the least-cost
     curve...186
7.5--Cost-effectiveness of alternative
     strategies for reducing sulfur
     deposition in the Adirondacks.................193
7.6--State-level distribution of SO_2
     emission reductions under S.768, as
     proposed and when targeted....................194

7.7--Distribution of utility SO_2 emissions
by start-up date and emission limit.............197
7.8--Projections of future utility SO_2
emissions under different retirement
assumptions......................................199
7.9--Emission reduction profiles for
alternative, age-dependent control
strategies.......................................201
7.10--The effects of a regional bubble
policy on power-plant emissions
rates after control..............................206
7.11--Emissions and costs of three bubble
options compared to current regulations..........208
7.12--Emissions and costs frontier for a
typical utility plant in Illinois
under the Aspin bill.............................212
11.1--Regional issues................................288
11.2--Cost-sharing...................................296

Preface and Acknowledgments

This edited collection on the technical and social dimensions of the acid rain debate grew out of our mutual feeling that the time had come to bring together the current research of several important scholars. Although we were aware of the number of books and monographs written in the last few years addressing the acid rain phenomenon, we believed that a policy-oriented book, taking as its essential point of departure the existing political stalemate over acid rain, was both necessary and possible. Happily, our invitations to researchers early in 1984 were quickly accepted, and the hard work began.

Every book is the product of the labors of many people who can never be fully rewarded or adequately recompensed for their efforts. Ours is no exception. We would, therefore, like to take some space to acknowledge the individuals who gave of their time and energies to help bring this work to fruition. First, our appreciation goes to each of the contributors who--in working with us on this volume--tolerated our periodic silences and then our frenetic calls and written requests with good cheer and graciousness. Then, too, we would like to thank Dean Birkenkamp, editorial director of the Special Studies series of Westview Press, for his willingness to proceed with the project on the basis of a brief prospectus, a hurried meeting in Washington, D.C., and a handful of early drafts of prospective chapters. No less important to bringing this work to completion was a timely grant to fund the typing of the camera-ready manuscript provided by Mike Crow of the Office of Research Stimulation and Interdisciplinary Projects of the Institute for Mining and Minerals Research (IMMR) at the University of Kentucky. The two women who patiently typed and retyped several drafts of various chapters--Pat Wade and Carol Hardy--have our thanks and appreciation. Kim Hayden, manuscript typist in the Department of Political Science at the University of Kentucky, has our deep gratitude both for assisting in the typing of several of the chapters and for working her magic on the printer to type the final version.

Special thanks go to Elizabeth Yanarella and Robin Ihara, who endured this book project from beginning to end with patience and good humor. Their unflagging support and steady affection were invaluable in the completion of this book. We would also like to give special recognition to Herbert G. Reid, whose friendship, counsel, and wide-ranging scholarship have been a source of intellectual inspiration and support for well over a decade. Finally, we wish to acknowledge the many scientists, scholars, environmental activists, politicians, and industry representatives whose dedication, energy, and abilities have provided a constant challenge to us in addressing the many difficult issues of the acid rain debate.

As with any jointly edited project, each of the co-editors would like to foist blame for any errors of fact or interpretation onto the other co-editor or, failing that, onto the individual contributors. In a more serious vein, unless explicitly stated otherwise, the views presented in the individual chapters of the book are those of the authors themselves and are not to be attributed to the institutions with which they may be affiliated.

--Ernest J. Yanarella
Lexington, Kentucky

--Randal H. Ihara
Arlington, Virginia

Introduction

A tangle of scientific and political issues has set the condition for the acid rain debate both in North America and in Europe. In the United States, this policy dispute has become a political stalemate at virtually every level of decisionmaking, pitting one constellation of political power against another. Meanwhile, members of the scientific community have seen their studies become political tokens in the debate while they themselves have become sometimes willing, sometimes reluctant, political actors in the controversy.

The following chapter by Randal H. Ihara introduces a number of political and technical issues. Taking as his point of departure the debate that emerged from congressional efforts to re-authorize the Clean Air Act, Ihara shows how this political controversy has generated "a kind of politics of risk assessment" stemming from competing regional views of the relative costs and benefits of acid rain mitigation. Ihara clearly illustrates the contrasting assessments of regional economic costs vs. regional environmental benefits, as well as the differing evaluations of the status of scientific knowledge about the many facets of acid deposition and its impacts.

Because scientific studies play such a prominent part in the acid rain debate, Ihara provides a summary of the latest research on acid deposition, particularly, "the relationship between emissions of SO_2 and NO_x and acid deposition, and the relationship between acid deposition and damage to aquatic resources and forests." While noting the gaps in scientific understanding of the phenomenon in the United States, he points to those findings and conclusions on which a reasonable consensus exists. His careful reading of key scientific studies and reports on the critical source-receptor issue--i.e., "the relationship between emissions from distant source regions...and the deposition of acidic compounds in sensitive receptor regions"--prompts him to emphasize the highly cautious and qualified conclusions of the 1983

National Research Council report. On the aquatic and terrestrial effects of acid deposition, Ihara's review of major research leads him to underscore the significant uncertainties about the complex interaction of acid deposition with aquatic and forest ecosystems.

Based on his experience as a professional staff member in the legislative policy process and his understanding of the science and politics of environmental issues, he predicts that a continuing policy impasse is likely "in the absence of the crystallization of scientific consensus." From Ihara's perspective, the policy logjam over acid rain control is unlikely to be overcome unless a new level of scientific understanding of this phenomenon is achieved--and perhaps until policy analysts come up with new ways of brokering the competing cost-benefit assessments and regional interests.

1
An Overview of the Acid Rain Debate: Politics, Science, and the Search for Consensus

Randal H. Ihara

INTRODUCTION

The acid rain debate has been a dominant issue in the environmental policy arena since 1981. For years, scientists and others have expressed concern over the adverse environmental effects of acid precipitation.[1] In response, the 96th Congress enacted the National Acid Precipitation Act of 1980, which established a ten-year national program of scientific research into the causes and consequences of acid precipitation. This legislation was an amendment adopted by the Senate during floor debate on the Energy Security Act--an ambitious energy policy initiative inspired by the Iranian oil crisis of 1979. Although the acid rain issue appeared to be overshadowed by energy policy concerns at the end of the last decade, it has become one of the more prominent, complex, and divisive policy issues of the 1980s. The acid rain issue has caused political tensions between the United States and Canada, divided the coal industry, pitted various regions of the nation against each other, and has caused divisions within both political parties.

In the 97th and 98th Congresses the Congressional hearing rooms became the principal battlefields for the acid rain policy debate, the terms of which were established as representatives of affected regions, industry, and the environmental and scientific communities clashed over the merits and deficiencies of the numerous legislative proposals relating to acid rain.[2] The issues and terms of the debate have, in turn, been further refracted through the media by reporters specializing in environmental affairs to the point that the acid rain issue is widely recognized by the general public as an environmental problem.[3]

Acid rain has become one of the major priorities on the nation's environmental policy agenda largely as a result of Congressional efforts to re-authorize the Clean Air Act. These efforts in the 97th, 98th and--as of this writing--the 99th Congresses have been frustrated by the political intractability of the acid rain issue. In 1982, and again in 1984, the Senate Committee on Environment and Public Works, chaired by Republican Senator Robert Stafford of Vermont, reported amendments to the Clean Air Act, including an acid rain program.[4] In this same period the House Subcommittee on Health and Environment, chaired by Representative Henry Waxman of California, struggled without success to report a bill.

The Senate Environment Committee program was intended to achieve significant reductions of acidic deposition in the eastern United States. The strategy called for a reduction in sulfur dioxide emissions of 8 million tons within a ten-year period in the eastern United States. The burden of achieving such reductions was primarily to be placed on electric utilities, the emissions of which exceed 1.2 lbs of sulfur dioxide per million Btus of energy input. This formula for allocating specific emission reduction targets essentially placed the burden of reduction on coal-fired utilities in midwestern and Appalachian states. Utilities required to reduce emissions would be allowed to meet their specific targets in the most economically efficient manner. For example, a utility would be allowed to achieve its reduction target by installing flue gas de-sulfurization equipment, so-called "scrubbers", or by switching to a low sulfur fuel, such as low-sulfur coal.[6]

The crux of the acid rain debate is whether or not the United States should adopt a regulatory strategy intended to achieve significant reductions of compounds which are considered to be precursors of acidic deposition (e.g., sulfur dioxide, nitrogen oxides) above and beyond the requirements of the Clean Air Act. The purpose of this chapter is to review some of the portentous political and technical issues fueling the acid rain debate in the United States. The first section will examine how this controversy has triggered a kind of politics of risk assessment pitting various regions of the country against one another. The next two sections examine the technical foundations of acid rain policy issues by summarizing the scientific status of research into those technical issues which bear on the control policy question.

COSTS, BENEFITS, AND THE POLITICS OF REGIONAL RISK ASSESSMENT

Proponents of a regulatory strategy, primarily representatives from northeastern states, argue that the weight of the scientific evidence clearly shows that acidic deposition is occurring in the northeastern

United States, as well as Canada and Western Europe, as a consequence of the chemical transformation of sulfur dioxide and nitrogen oxides emitted by electric utilities and other industrial facilities which use fossil fuels. These emissions are carried high into the atmosphere where they may be transported by prevailing winds hundreds of miles from their point of origin, transformed into acidic materials in clouds, and deposited on the earth's surface as rain, snow or fog.

When these acid materials are deposited on sensitive aquatic or terrestrial ecosystems, damage is caused, including reduced productivity of forests and croplands. Hundreds of lakes in the Adirondack Mountains of New York no longer support fish as a result of acid rain. Thousands more lakes which are sensitive to acid rain through the eastern United States are threatened with the same fate. Furthermore, the deposition of acidic materials can corrode and dissolve the surfaces of buildings and public monuments. Finally, human health may be jeopardized by the corrosion of the copper and lead pipes which are used in municipal water systems. The dissolved metals can then find their way into drinking water supplies, thus posing a threat to public health.

Proponents of acid rain legislation further argue that the economic costs of acid rain are substantial. For example, the impact of damage from acid rain on current economic activity in the eastern United States has been estimated to be about $5 billion per year. It has also been estimated that the cumulative cost of damage to fisheries and crops from acid deposition may be as high as $15 to $25 billion by the year 2000.[8] In view of these considerations, the proponents of acid rain legislation contend that acid rain damage to the Northeast can only be mitigated through substantial reductions in the emissions of sulfur dioxide and nitrogen oxides from coal-fired powerplants in the Midwest and Appalachia.

Opponents of a regulatory strategy--primarily representatives of the industrial midwestern and Appalachian states--argue that such an approach is "unreasonable in its costs, unfair in its burdens, and...may not achieve its objectives."[9] It is recognized that estimates of the costs of adopting a regulatory strategy vary widely, depending upon the assumptions employed regarding the extent to which utilities will choose to meet emission reduction requirements through the use of conventional emission control technology (e.g., "scrubbers") or by switching to low-sulfur coal. The magnitude of reductions required, as well as other factors, may also affect cost estimates.[10] However, even with these considerations in mind, the costs are large. It has been estimated that, in 1995, cumulative capital costs in the states affected by an acid rain regulatory program would be in the range of $7.2 to $17.8 billion (1980 dollars), depending on the extent of "scrubbing" relative to fuel-switching by electric utilities. Other estimates are significantly higher.[11]

Furthermore, increased capital costs and increased operations and maintenance costs will be reflected in substantial increases in electric utility rates for residential and industrial customers in the Midwest and Appalachia.[12] Alternatively, to the extent that utilities choose to switch to low-sulfur coal rather than to install scrubbers to meet emission reduction requirements, high-sulfur coal-producing regions in the Midwest and Appalachia would suffer substantial unemployment impacts. According to one estimate, total job losses in the high-sulfur coal-producing counties of Illinois, Indiana, Ohio, Pennsylvania, Kentucky and West Virginia would be from 42,000 to 57,000.[13] Traditional coal markets would be disrupted, as utilities choose to switch to low-sulfur coal. This would cause a reduction in demand for high-sulfur coal. High-sulfur coal production is primarily located in the Midwest and northern Appalachia. Low-sulfur coal-producing regions--mainly located in the western United States, as well as parts of southeastern West Virginia, parts of eastern Kentucky and southwest Virginia--would experience substantial increases in demand.

Opponents of a regulatory strategy also contend that midwestern and Appalachian states would be unfairly penalized by formulas which have been proposed for allocating emission reduction targets among the affected states. For example, they point to the legislation reported by the Senate Environment Committee in 1982, which would have required larger proportionate reductions from the midwestern and Appalachian states than from those in the Northeast. Kentucky and New York both were reported to have had similar levels of SO_2 emissions in 1980. Kentucky, however, would have been required to reduce emissions by almost 63%, while New York would have been required to achieve a reduction of 21%. In 1980, Mississippi had SO_2 emissions which were lower than those of New Jersey. Yet Mississippi would have been required to reduce emissions by 11%.[14]

Since the regional economic stakes are perceived to be so high, the scientific underpinnings of a program purporting to mitigate acid deposition become a crucial issue. It is not surprising then that one of the central issues of the debate is the status of scientific understanding of the causes and consequences of acid deposition. Proponents of a regulatory program contend that the weight of the scientific evidence to date is sufficient for Congressional action. While they grant that there are scientific uncertainties yet to be addressed, "that uncertainty must be weighted against the risks associated with inaction."[15] Senator George Mitchell, a leading proponent of a regulatory approach to the acid rain problem, defined the issue well:

> I emphasize the high risk of inaction. I recognize the fact that some questions about acid precipitation remain to be answered. But I also believe that evidence to support a meaningful sulfur reduction strategy is stronger and more complete than in most legislative debates...In my view, those who argue for postponing action now are demanding adherence to a standard of certainty that has never been met in the past and can never be met in the future.[16]

At issue is the assessment of the risks of action or inaction on the acid rain problem. The terms of this assessment are both economic and environmental. Thus, Senator Robert Stafford of Vermont contends that "the risk is quite literally the economy of an entire region, a major region of the United States [i.e., the Northeast]."[17] Senator Mitchell has argued that "the issue of acid rain is first and foremost an environmental problem...We are confronted with a phenonmenon that is causing irreversible damage to our natural resource base."[18]

Opponents of acid rain regulatory programs assess the risks in similar terms, though with different conclusions. They point to the risks to the economies of midwestern and Appalachian states if a regulatory program is enacted.[19] In addition, they question whether there is a need for immediate action. "There is an acid rain problem," Senator Richard Lugar of Indiana has argued, "but no acid rain crisis."[20]

The issue of whether or not scientific understanding is adequate for policy purposes is implicit in these opposing assessments of risk. Representatives of those states which are expected to bear the adverse economic impacts are demanding far more stringent a standard of scientific proof than those advocating the adoption of a regulatory program. In other words, those who are concerned that they will bear the economic costs are imposing a far greater burden of proof on the scientific community than those who would apparently enjoy the benefits. As a result, opponents of an acid rain control program are demanding that scientific consensus on the causes and

FIGURE 1.1
Annual mean value of pH[a]

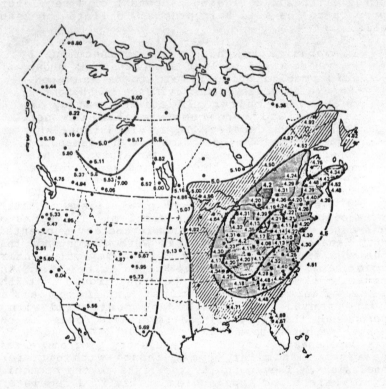

[a]Weighted by amount of precipitation in the United States and Canada for 1980.

Source: U.S.-Canada Work Group 2. *Atmospheric Science and Analysis: Final Report.* Washington, D.C.: U.S. Environmental Protection Agency, 1982.

consequences of acid rain be a prerequisite for a political consensus, while proponents claim that a sufficient scientific consensus already exists on which to base a political consensus.[21]

Support for a regulatory strategy to address the acid rain problem is largely driven by concerns about the environmental and other effects of acid rain. These concerns have been fueled by--and in turn, have helped generate--a large, highly complex, and ever-expanding body of scientific literature regarding the causes and consequences of acid rain. This literature is summarized below in order to provide an overview of the status of scientific understanding about key aspects of the acid rain issue--and, in particular, the relationship between emissions of SO_2 and NO_x and acid deposition, and the relationship between acid deposition and damage to aquatic resources and forests.

THE SCIENTIFIC STATUS OF ACID RAIN POLICY ISSUES: EMISSIONS AND ACID DEPOSITION

As indicated by Figure 1.1, the pH of rainfall is lower in eastern North America, including the northeastern United States (with a pH of 4.2) and southeastern Canada, than other regions of the continent (with a pH of about 5.0). There is a scientific consensus that anthropogenic emissions of sulfur dioxide and nitrogen oxides have increased the levels of precipitation acidity over eastern North America.[22]

Many efforts have been made to establish long-term historical trends in the pH of precipitation over the eastern United States through a reconstruction of disparate data sets; however, scientific control procedures and collection techniques for these data have been varied and inconsistent.[23] Different techniques for determining pH values were used from some data, and in some cases the techniques used for a particular data set at one point in time were later changed.[24] Thus, reliable long-term historical data are not available to present a detailed picture of trends in precipitation acidity in the United States.[25] Although it is difficult to establish any long-term trends on a reliable basis for the eastern United States, there is general scientific agreement that "precipitation is definitely acidic over this region, and is probably more acidic than expected from natural baseline conditions."[26]

Emissions of sulfur dioxide are thought to play a major role in the acidity of precipitation in eastern North America. Sulfur dioxide, when transformed into H_2SO_4, accounts for a large share of the hydrogen ion

TABLE 1.1
National U.S. current and projected
SO_2 and NO_x emissions (Tg/yr)

Source category	Current 1980		Projected 1990		Projected 2000	
	SO_2	NO_x	SO_2	NO_x	SO_2	NO_x
1. Electric utilities	15.0	5.6	15.9	7.2	16.2	8.7
2. Industrial boilers and process heaters	2.4	3.5	3.4	3.0	6.5	4.0
3. Nonferrous smelters	1.4		0.5		0.5	
4. Residential/commercial	0.8	0.7	1.0	0.7	0.9	0.6
5. Other industrial processes	2.9	0.7	1.2	0.8	1.5	1.1
6. Transportation	0.8	8.5	0.8	7.8	1.0	9.7
TOTALS	24.1	19.0	22.8	19.5	26.6	24.1

Source: U.S.-Canada Work Group IIIB.
<u>Emissions, Costs</u> <u>and</u> <u>Engineering</u>
<u>Assessment</u>: <u>Final</u> <u>Report</u>. Washington,
D.C.: U.S. EPA, 1982.

deposition in eastern North America. As shown in Table 1.1, electric utilities accounted for about 62% of the total U.S. emissions of sulfur dioxide in 1980. By the year 2000, powerplants are expected to account for about 61% of the total. In relative terms, powerplant emissions of SO_2 are projected to be stable to the end of the century, and then to decline after the year 2000. It should be emphasized, however, that these projections are highly uncertain, since they are based on several arguable assumptions--including assumptions about the rate at which new coal-fired capacity (the emissions of which are governed by NSPS) will come on line in the near future, and the rate at which older, pre-NSPS capacity is retired. Utility economics are in the midst of dramatic changes which will certainly affect the validity of such assumptions.[27]

In 1980, the primary sources of NO_x were the transportation sector (45%) and powerplants (29%). However, by the year 2000, NO_x emissions from powerplants are projected to increase to 36% of total U.S. emissions of NO_x. The transportation sector is projected to remain the leading generator of NO_x (40%). However, note that utility NO_x emissions are projected to increase significantly, while the transportation sector's share of total NO_x emissions is expected to decline.

The direction of utility emission trends has significant policy implications. If emissions should remain stable in the mid-term future and then begin to decline after the turn of the century, then current policy (i.e., the Clean Air Act) could be viewed as an effective, long-term program for controlling acid rain. Such a scenario would allow for a more orderly, less disruptive development of coal markets. On the other hand, if emissions should increase perhaps as a result of the changes in utility economics and despite implementation of the Clean Air Act, affirmative regulatory action may be appropriate in order to bring down future emission levels.

The critical issue for governmental policy is the nature of the relationship between emissions from distant source regions (such as the Midwest) and the deposition of acidic compounds in sensitive receptor regions (such as the Northeast). That is, is that relationship such that a reduction of SO_2 in the source region will produce a commensurate reduction in deposition in the receptor region?[28] One major difficulty in answering this question is that the compounds emitted from a distant source region and the compounds deposited in a receptor region are chemically different. For example, SO_2 is emitted from a source region, but the compound deposited in the receptor region is H_2SO_4. Establishing the relationship quantitatively between emissions and deposition (i.e., the source-receptor relationship) requires an understanding of meteorological conditions relative to a number of critical processes--including atmospheric transport, the transformation of acid precursor compounds into acids, and deposition of these acids as well as sulfates and nitrates in dry and wet form.

Currently, scientific understanding of these processes is incomplete.[29] After an extensive review of the status of scientific research into the relationship between emissions and deposition, the National Research Council concluded that "there is no evidence that the relationship between emissions and deposition in northeastern North America is substantially non-linear when averaged over a period of a year and over dimensions of the order of a million square kilometers." Thus, "if the emissions of sulfur dioxide from <u>all sources</u> in this region were reduced by the <u>same fraction</u>, the result would be a corresponding fractional reduction in deposition."[30] Clearly, the NRC's statement is heavily qualified--as was its entire report.

The council report indicates that its conclusion on the linearity of the source-receptor relationship "is clouded" by uncertainties in three areas. First, the data base upon which the conclusion relied is very limited. There are only ten stations in North America from which long-term data could be drawn. Of those ten, the council considered the data from only one (the U.S. Forest Service experimental station at Hubbard Brook) to be "reasonably reliable." Secondly, the council noted "uncertainties in our detailed understanding" of the various processes linking emissions and deposition. Finally, the council's confidence in its conclusion is further limited by "the unknown influences of natural variability in the composition and occurrence of precipitation, imprecision in sampling and analysis, and uncertainties in estimation of emissions."[31]

The council did conclude that evidence exists to support the thesis that acid precursor compounds are carried long distances from source regions. On the other hand, it also pointed out that the deposition at any given receptor area would include compounds from <u>both</u> distant and local sources. Pollutants transported from distant sources and those from local sources are thoroughly mixed over a wide geographical area as a result of various atmospheric processes. Thus, the council could not "in general determine the relative importance for the net deposition of acids in specific locations of long-range transport from distant sources or more direct influences of local sources."[32]

There may be distinct regions in eastern North

America where the contributions of local emission sources (i.e., sources 500 km or less from a receptor area) and distant SO_2 sources (i.e., more than 500 km) may be markedly different. Acidic deposition in the upper Ohio Valley may be dominated by local sources, while in upper New England and parts of eastern Canada distant sources may be responsible for a large share of the acidic deposition. In more intermediate regions, including the Adirondacks of New York, "contributions to total acidic deposition from near and far sources may be comparable, considered on a regional and summer average basis."[33]

Despite general agreement on these points, the status of scientific understanding of the processes involved in the transport and transformation of precursor compounds to acids severely limits the capability of "identifying specific sources responsible for wet deposition at a given receptor site."[34]

ACID DEPOSITION AND NATURAL RESOURCE DAMAGE: AQUATIC AND TERRESTRIAL EFFECTS

Acid Deposition and Aquatic Resources.

The acidification of lakes and streams in the eastern United States and the accompanying biological damage to fish and other aquatic biota have been widely reported for the past ten years. Yet, there is considerable scientific uncertainty regarding the extent of damage, the resources at risk, the rate at which damage is occurring (or will occur), and the reversibility or irreversibility of the damage. The principal difficulty is that the "aquatic response to acidic deposition is largely controlled by processes within the terrestrial ecosystem."[35] According to recent scientific studies,

> [a]ssessing causal relationships remains difficult...because effects of acidic deposition on any one component of the terrestrial-wetland-aquatic systems depend not only on the composition of the atmospheric deposition but also on the effect of the atmospheric deposition on every system upstream from the component of interest. Composition of aquatic systems results, moreover, from biological processes in addition to chemical and physical processes; thus assessing results of acidification on all three processes is required. Our knowledge of past, current, and future acidification, of critical

processes that control acidification, and of
the degree of permanency of chemical and
biological effects remains incomplete and
subject to debate.[36]

The evidence to date suggests that surface waters in
the northeastern United States have been affected by
acid inputs.[37] There is, however, considerable debate
regarding the contribution of long-range transport of
acidic materials relative to other resources of acidic
inputs and their effects on surface waters in this
region.

In many regions of North America the geological
and physical characteristics of aquatic systems and
their terrestrial surroundings are such that there is
little capacity for neutralizing acid inputs (e.g.,
thin soils, granite or metamorphic bedrock) from any
source.[38] The sensitivity of aquatic systems to
acidification is dependent upon the physical and
geographical characteristics of the lake or stream, the
composition of the deposition (i.e., the amount of
sulfate, nitrate, base cations, etc.), the total rate
of acid loading (the sum of both wet and dry
deposition), and the temporal distribution of the
deposition (e.g., seasonal, episodic, long-term).[39]
The interaction of these factors may be highly variable
between, and even within, geographic regions. Thus,
the sensitivity of aquatic systems is also highly
variable and cannot be accurately characterized simply
on the basis of a single indicator, such as bedrock
geology.[40]

Regions with aquatic systems thought to be
sensitive to acidification include "eastern Canada and
New England, parts of the Allegheny, Smoky, and Rocky
Mountains, the northwestern and north central United
States, and the south and east coasts of the United
States." Considerable work remains to be done to
determine levels of alkalinity and the degree of
sensitivity of individual aquatic systems.[41]

Empirical evidence suggests that acidic deposition
has contributed to, or has caused, the substantial
reduction of alkalinity in sensitive North American
aquatic systems.[42] Still, it appears that lakes where
alkalinity may have been exhausted by acid deposition
(so-called "dead lakes") represent a small percentage
of the surface waters of the eastern United States,[43]
and may be limited to relatively small, high elevation
lakes in the Adirondack Mountains of New York and upper
New England. Without an extensive survey of aquatic
systems, the extent of damage due to acidification
remains unclear.

The picture is further complicated by recent
studies of the United States Geological Survey

regarding trends in the acidity of surface waters in the United States and Canada. In 1982, the U.S.G.S. published a study of trends in the acidity of precipitation and surface waters in New York. The report found that between 1965 and 1978 there was little change in the pH of precipitation within New York state as a whole. However, since 1965, the eastern part of the state has experienced an increase in the pH of precipitation; western New York by contrast has experienced a decline in precipitation pH during the same period.

The U.S.G.S. report also found that sulfate concentrations in streams had <u>declined</u> an average of 1 to 4 percent per year, reflecting a similar trend in the sulfate concentrations in New York's precipitation.[44] Trends in stream acidity due to precipitation, however, were difficult to discern due to "the greater impact of agriculture and urbanization." Nonetheless, the Geological Survey found that concentrations of sulfate in streams decreased 1 to 4 percent per year, which suggests a decrease in levels of deposition.

Scientific research has also indicated that, while sulfate concentrations have declined in the Northeast over time, concentrations of nitrates may be increasing. The net effect is that no long-term trend in precipitation pH is discernible.[45] The increased concentrations of nitrates also suggest that NO_x may play a much more important role as a major contributor to the problem of acid deposition than has been generally recognized by policymakers.[46]

Other scientific evidence suggests that acidification of lakes and streams in the eastern United States may have "occurred before the mid- to late 1960's," and may have stabilized at current levels of deposition. In addition, there is some evidence that aquatic systems in the western United States may be undergoing acidification, and that local sources of acidity may play the dominant role.[47]

The critical issue with respect to acid desposition and aquatic systems is the rate at which acidification is occurring. There are two hypotheses about the process of acidification and its pace. The first has been referred to as the "delayed response" hypothesis, and the second as the "direct response" hypothesis. These hypotheses relate to the <u>capacity</u> of a watershed to neutralize acid inputs, and the <u>rate</u> at which a watershed can supply base cations to neutralize acid. That is, a watershed's sensitivity to acidity can be "capacity limited" or "rate limited." (It is also worth noting that the two hypotheses are not mutually exclusive. Indeed, capacity limited watershed and rate limited watersheds may be part of a continuum in different regions of the nation.)

According to the delayed response hypothesis, acidification occurs over a long period of time. A lake or stream may have a high capacity to retain sulfates deposited from the atmosphere in the surrounding soil, but may have a limited capacity to supply base cations to neutralize acidic anions leaving the soil and flowing into the water. As the capacity to supply base cations diminishes from continued acid input, alkalinity will decrease as more and more unneutralized acid enters the lake or stream. On this view, damage to capacity limited aquatic systems is cumulative.

In terms of its policy implications, then, the delayed response hypothesis suggests that, at current levels of deposition, the neutralizing capacity of more and more watersheds will be depleted.[48] It also suggests that there may be a threshold point at which a lake's neutralizing capacity is exhausted. Once that threshold is passed, the lake will suddenly be "dead." Thus, while the number of dead lakes is currently relatively few, hundreds or thousands more may suddenly appear as the neutralizing capacity is exhausted by continued acid deposition.

The direct response interpretation hypothesizes that acidification is a more immediate response to acid deposition. Lakes or streams, such as those found in the northeast, with a low capacity to retain sulfate and a low ability to release base cations, will tend to become acidified relatively quickly because the sensitivity of the lake or stream is determined by the rate at which it can neutralize incoming acids.[49]

The implication of this hypothesis is that the "acidification which has occurred in our lakes and streams is all that will occur unless future levels of acidic deposition increase."[50] In policy terms, this suggests that aquatic systems may have reached a point of equilibrium; and in some instances, acidification may be reversing in response to general declines in SO_2 emissions and deposition.

There is general agreement that in "lakes and streams impacted by acid deposition, where sulfate concentrations are still increasing, the alkalinity will continue to decrease." In those lakes and streams where sulfate concentrations appear to be stable, "there is disagreement...as to whether future alkalinity decreases will be gradual over the period of centuries or perceptible over years to decades."[51]

The policy implications of these issues are clear. If, at current levels of deposition, aquatic systems have reached an equilibrium point, or if alkalinity decreases will be gradual over a long period of time, then the necessity for immediate affirmative action to reduce deposition levels is diminished considerably. If, on the other hand, alkalinity decreases are perceptible within a few years, then immediate action to reduce deposition levels might be an appropriate policy objective.

Acid Deposition and Aquatic Biology.

Concern for acidification of aquatic systems is driven by the suspected biological impacts of acidification. The loss of fish and other aquatic organisms and the association of such loss with acidification have been reported in the scientific literature and widely covered by the media.[52]

The loss of fish as a consequence of acidification has been documented for particularly sensitive aquatic systems in five regions in North America and Scandinavia: the LaCloche Mountain region of Ontario, Nova Scotia, southern Norway, and southern Sweden. The loss of fish populations has been documented in 180 lakes in the Adirondack Mountains of New York, the only region in this country where such loss has been documented. Indeed, while acidification of these lakes "probably contributed to the disappearance of fish for at least some surface waters," a direct link between acidification and fish loss in the Adirondacks has not been substantiated. In addition, the extent of such loss cannot be "satisfactorily evaluated at this time."[53] Although the relationship between the acidity of a lake or stream and fish population response is complex and variable, there is general scientific agreement that:

> [b]iological effects due to acidification occur for a few species near pH 6.0. Because the biological response to acidification is a gradual one, continuing pH declines below pH 6.0 will result in escalating biological changes, with many species adversely affected in the range of pH 5.0 to 5.5...In addition, episodic depressions down to pH 4.4 to 4.9 often occur in low alkalinity waters during periods of snowmelt and heavy rainfall and can affect systems with a pH as high as 7.0. These pH levels...will have significant harmful effects on aquatic organisms. In waters where pH values average below 5.0, most fish species, virtually all molluscs, and many groups of benthic invertibrates will be elminated. Increased aluminum concentrations may eliminate fish species otherwise tolerant of low pH.[54]

Acid Deposition and Forest Effects

Concern in Europe and the United States has grown over recent evidence of forest decline of unprecedented magnitude, including a variety of tree species in Central Europe, the United States, and Canada. Because the damage is occurring in regions which receive elevated levels of air pollutants, it is assumed that there is a strong and direct relationship between air pollution and forest decline.[55] Reports of widespread damage to forests are most dramatic in West Germany where a 1982 forest inventory indicated that 8% of the forest area had been damaged. One year later, a repetition of that inventory revealed that 34% of the West German forest area was damaged. In 1984, it was estimated that 50% of the forest area was damaged.[56] Other reports have indicated damage in high-elevation forests in France, Norway, Italy, Austria, Poland, Yugoslavia, and Czechoslovakia.[57]

In Europe, a variety of tree species have been affected, including white fir, Scots pine, Norway spruce, and beech.[58] In the United States, damage to various tree species (e.g., red spruce, birch, pitch pine, loblolly pine, and short leaf pine) has been observed in forests at Camel's Hump (Vermont), Hubbard Brook (New Hampshire), and Mount Mitchell (North Carolina). Of particular concern are reports that "a wide variety of forests covering a broad area have experienced substantial losses of spruce. Red spruce in the southern Appalachians appears to be unaffected, while from the Catskill Mountains of southern New York northward, considerable dieback and mortality are noted."[59]

Three types of forest decline have been identified. The first involves widespread, rapidly developing, and synchronous (1-5 years) decrease in the diameter growth of affected trees. In this type, few other visible symptoms are apparent. The second type is similar to the first, except that the decrease in diameter growth is followed by other visible symptoms (e.g., yellowing of needles, crown dieback) which often lead to the death of the tree. The third type of forest decline involves progressively developing decrease in diameter growth over a period of 3 to 15 years. Other visible symptoms occur which can lead to the death of the trees.[60] All three types of decline may be found in the United States.

It is generally agreed that the decline of forests and the decline of individual trees may be the result of interactive effects of air pollutants with other environmental stress factors.[61] However, specific mechanisms of damage have not yet been established. Indeed, there is considerable uncertainty about the relationships--if any--between acid deposition and forest decline. At the present time, there are a number of major hypotheses regarding those relationships:

Acidification/Aluminum Toxicity: This hypothesis suggests that acid deposition accelerates the naturally occurring acid-producing processes in the soil. The increased production of acid frees aluminum ions bound to other compounds in the soil. The resultant increase in the concentration of aluminum ions leads to necrosis of the fine root hairs of a tree, which, in turn, constrains the ability of the tree to absorb moisture and nutrients from the soil. Since the tree is unable to absorb moisture, it becomes dehydrated and particularly vulnerable during periods of drought.[62]

Ozone: This hypothesis suggests that there is a relationship between the high levels of photochemical oxidants, particularly ozone, which prevail in the eastern United States at phytotoxic levels, and laboratory and field observations of damage to conifers and broadleaf trees.[63]

General Stress: According to this hypothesis, net photosynthesis is reduced as a result of air pollution and the deposition of toxic and other materials (e.g., heavy metals). The toxic substances accumulate in roots and affect the development of fine root hairs. This can result in symptoms of foliar decline. In this condition, the tree becomes vulnerable to a variety of naturally occurring stress factors, such as drought and attack by secondary or biotic pathogens.[64]

Excess Nutrient (or Excess Nitrogen): This hypothesis suggests that atmospheric concentrations of nitrogen compounds are contributing to forest decline. Nitrogen is an essential nutrient, but when supplied in excess of tree requirements, it can have a number of adverse consequences, including increased susceptibility of the tree to frost, root disease, or nutrient leaching from the surrounding soil.[65]

While each of these hypotheses addresses some observed aspects of forest decline in North America and Europe, no one hypothesis is adequate for all the observed effects. For the present,

> there is no proof that acidic deposition is currently limiting growth of forests in either Europe or the United States. From field studies of mature forest trees it is apparent that altered growth patterns of principally coniferous species examined to date have occurred in recent decades in many areas of the northeastern United States and

in some areas of Europe with high atmospheric deposition levels. Recent increases in mortality of red spruce in the northeastern United states and Norway spruce and beech in Europe add further to the concern that forests are undergoing significant adverse change; however, no clear link has been established between these changes and anthropogenic pollutants, particularly acidic rainfall.[66] The high degree of complexity and variability of forest ecosystems limits the extent to which generalized conclusions can be drawn.

CONCLUSION

The stock of scientific knowledge on acid deposition and its environmental effects is evolving rapidly, particularly with respect to lakes and forests.[67] Clearly, scientific understanding remains incomplete or in its early stages of development in many areas--such as the causal relationships between acid deposition and damage to lakes and forests, the rates at which damage is occurring, and the extent of damage caused by acid deposition.[68]

These issues are of relevance to policymakers, but there are disagreements about whether "enough" is known for affirmative action. As research continues one may reasonably expect some further scientific clarification. Nevertheless, the question of whether we know "enough" is a matter of political, not scientific judgment. At this point, science can shed light on many issues, but it is one consideration among many in the policy debate. Costs and benefits must be weighed against risks and probabilities. There must be judgments about economic and ecological consequences. For example, proponents of acid rain legislation raise the spectre of irreversible ecological damage if current levels of deposition are allowed to continue. However, opponents raise the issue of whether the adverse consequences of such legislation for the economies of the Midwest and Appalachia are any less irreversible.

Congress, and especially the Senate, is an institution which operates on the basis of accommodation and consensus. The regional and other cleavages on the acid rain issue, as well as the on-going debate about the science of acid rain, do not lend themselves to either. The key policy issues include whether to impose emission controls. If so, which emissions should be controlled and how much control is necessary? Should all sources of emissions be controlled? Should a control program be national in scope, or regional? Who should pay the costs of controls, and how should those costs be distributed? In the absence of the crystallization of scientific consensus around the issues highlighted above, the prospect for a resolution of any of these policy issues by the Congress in the near future is exceedingly problematic.

NOTES

   1. A useful historical perspective may be found in Ellis B. Cowling, "Acid Precipitation in Historical Perspective," **Environmental Science and Technology,** Vol. 16 (February 1982), pp. 110A-123A.
   2. As of the end of 1983 there had been almost 20 acid rain bills introduced in the Senate and House, most of which were variations on a theme. An analysis of the major bills can be found in D.G. Streets, L.A. Conley, L.D. Carter, and J.E. Vernte, **Analysis of Proposed Legislation to Control Acid Rain** (Argonne, Ill.: Argonne National Laboratory, January, 1983). An analysis of the Senate Environment Committee bill is: ICF, Inc. **Analysis of a Senate Emission Reduction Bill (S. 3041)** (Washington, D.C.: ICF, Inc., February 1983).
   3. See the compilation of editorials and public opinion polls on acid rain in Congressional Research Service, **Overview: Acid Rain** [VU 84016] (Washington, D.C.: CRS, 1984).
   4. The bills reported by the Senate Environment Committee in both 1982 (S. 3041) and 1984 (S. 768) were very similar. Consequently, for purposes of discussion, both bills will be treated as "the Senate bill." However, the 1984 bill will be used as the basis for the description of Senate provisions. The amendments were never brought to the Senate floor for a vote, largely due to their highly controversial nature. In the House, an acid rain program included in amendments to the Clean Air Act was killed in a narrow 10-9 vote by the House Subcommittee on Health and Environment in 1984. The legislation was principally crafted by subcommittee chairman Henry Waxman (D-Calif.) and Gerry Sikorski (D-Minn.). The bill was considered to be a promising compromise proposal which attempted to address the concerns of high-sulfur coal regions.
   5. The Sikorski-Waxman proposal (H.R. 3400) required that the 50 utility plants which are the largest emitters of SO_2 install scrubbers to reduce emissions of SO_2 by 7 million tons by 1990. The 48 contiguous states would be required to reduce emissions

of SO_2 an additional 3 million tons on a proportionate basis. The states would decide how to achieve these reductions. The bill would impose a national fee of 1 mill per kilowatt hour on all non-nuclear electricity generation. A trust fund would be established to reimburse utilities for 90 percent of the capital costs of installing emission control equipment. NO_x emissions would be reduced 1.5 million tons by 1995 through a tightening of the NSPS for NO_x. Tighter NO_x emissions standards would be imposed on truck emissions to achieve an additional 2 to 3 million tons of NO_x reduction by 1995. This was an important proposal, for it was the first to require the use of emission control technology, and to propose a cost-sharing program to help defray the capital costs of installing such technology. It was the first to propose a national fee and trust fund to fund a portion of the capital costs of the program. Finally, it was the first proposal explicitly to recognize the role of NO_x in the acid rain problem and to propose a program to address NO_x emissions. For an analysis of the costs and economic impacts of H.R. 3400, see ICF, Inc., **Analysis of the Waxman-Sikorski Sulfur Dioxide Emission Reduction Bill (H.R. 3400)** (Washington, D.C.: ICF, Inc., April 1984).

6. For a description of this strategy, and a discussion of its rationale, see **Clean Air Act Amendments of 1982**, report of the Committee on Environment and Public Works, Senate [Report 97-666] (Washington, D.C.: U.S.G.P.O., November 15, 1982), pp. 52-73; and **Clean Air Act Amendments of 1984**, report of the Committee on Environment and Public Works, Senate [Report 98-426] (Washington, D.C.: U.S.G.P.O., May 3, 1984), pp. 49-68. See, also, David G. Hawkins, "Controlling Acid Rain: A Modest Proposal," **The Amicus Journal**, Vol. 4 (Winter, 1983), pp. 12-13.

7. For a summary presentation of these points, see Robert A. Boyle and Alexander Boyle, "Acid Rain," **The Amicus Journal**, op. cit., pp. 22-37. See also the comments of Senator Robert Stafford, in **Acid Rain: A Technical Inquiry**, hearings before the Committee on Environment and Public Works, U.S. Senate [Serial 97-H53] (Washington, D.C.: U.S.G.P.O., May 25 and 27, 1982), pp. 243-244; statement of Senator George Mitchell, in **Acid Precipitation and the Use of Fossil Fuels**, hearings before the Committee on Energy and Natural Resources, U.S. Senate [Publication No. 97-87] (Washington, D.C.: U.S.G.P.O., August 19, 1982), pp. 129-131.

8. Comments of Senator George Mitchell in **Acid Rain: A Technical Inquiry**, pp. 247-248. A brief summary of selected cost estimates is contained in Congressional Research Service, **Acid Rain: A Survey of Data and Current Analyses**, a report prepared for the

Subcommittee on Health and the Environment of the Committee on Energy and Commerce, U.S. House of Representatives [Committee Print 98-X] (Washington, D.C.: U.S.G.P.O., May 1984), pp. 62-66.

9. Comments of Senator Richard Lugar, in **Acid Precipitation and the Use of Fossil Fuels**, p. 157.

10. See, Congressional Research Service, **Acid Rain: A Survey of Data and Current Analyses**, pp. 66-71.

11. See, Energy Information Administration, **Impacts of the Proposed Clean Air Act Amendments of 1982 on the Coal and Electric Utility Industries** (Washington, D.C.: U.S.G.P.O., June 1983), p. ix. See also, Larry Parker, "Summary and Analyses of Technical Hearings on Costs of Acid Rain Bills." Prepared at the Request of the Committee on Environment and Public Works, U.S. Senate (July 26, 1982).

12. The impact of acid rain legislation on electric utility rates is highly uncertain. Estimates generally show an increase, although the magnitude of such increases vary widely depending upon a number of factors. ICF, Inc. has estimated that electric rates in the eastern U.S. would increase by an average of about 2.5% in 1995. The highest any state would experience would be about 9%. See ICF, Inc., **Analysis of a Senate Emission Reduction Bill (S. 3041)**, pp. S-18. Estimates by electric utilities tend to be higher. For example, Associated Electric Cooperatives, Inc., an electric power cooperative in Missouri, estimated rate increases of 20-30% in 1988. Ohio Edison estimated increases of 40%. See Associated Electric Cooperatives, Inc., "Estimated Economic Impact of Proposed Acid Deposition Regulations on Associated Electric Cooperatives, Inc.," in **Acid Rain: A Technical Inquiry**, p. 265; "Presentation of Charles V. Runyon, Manager, Environmental Special Projects Department, Ohio Edison," **Ibid.**, p. 699; and U.S. Department of Energy, **Costs to Reduce Sulfur Dioxide Emissions** (Washington, D.C.: U.S.G.P.O., March 1983), p.25. Rate increases were estimated by the Department of Energy to be in a range of 4% to 26%.

13. Larry Parker, **Mitigating Acid Rain: Implications for High-Sulfur Coal Regions** (Washington, D.C.: Congressional Research Service, May 16, 1983). The legislative history is replete with estimates of the impacts of acid rain legislation on electric utility rates, utility capital expenditures, coal production, and employment. See for example, ICF, Inc., "Summary of Preliminary Analysis of the Costs and Coal Production Impacts of the Mitchell Bill," in **Acid Precipitation and the Use of Fossil Fuels**, pp. 60-75; testimony of Larry Reynolds, United Mine Workers of America, "Economic and Employment Impacts of Proposed

Acid Rain Legislation," **Ibid.**, pp. 77-90. National Economic Research Associates, Inc., **A Report on Results from the Edison Electric Institute Study of the Impacts of the Senate Committee on Environment and Public Works Bill on Acid Rain Legislation (S. 768)** (Washington, D.C.: NERA, Inc., June 20, 1983). A more recent, and more comprehensive, analysis is **Economic Impacts of Alternative Acid Rain Control Strategies** (Washington, D.C.: Arthur D. Little, Inc., January 1985). The state of Illinois is one example of the dramatic economic impacts of acid rain legislation. It has been estimated that as the result of an 8 million ton SO_2 reduction program, which employed a least-cost approach, Illinois coal production would decline 40% from 1980 levels. Coal industry unemployment would be about 21 percent, with a loss of about $200 million in personal income. See Larry Parker, "Impact of Proposed Acid Rain Legislation on the Illinois Coal Industry," prepared for the Senate Subcommittee on Energy, Nuclear Proliferation, and Government Processes, Senate Governmental Affairs Committee. (Washington, D.C.: Congressional Research Service, February 21, 1983).

14. See statement of Senator Wendell Ford in **Acid Precipitation and the Use of Fossil Fuels**, p. 115. The figures for Mississippi and New Jersey were calculated from **Acid Rain: A Technical Inquiry**, table 4, p. 38.

15. Comments of Senator Stafford, in **Acid Rain: A Technical Inquiry**, p. 2.

16. Comments of Senator Mitchell, **Acid Precipitation and the Use of Fossil Fuels**, p. 124.

17. Senator Stafford, in **Acid Rain: A Technical Inquiry**, p. 2.

18. Senator Mitchell in **Acid Precipitation and the Use of Fossil Fuels**, pp. 127, 131.

19. As Senator Robert C. Byrd, a leading opponent of the regulatory approach to the acid rain problem, has suggested: "my position...recognizes the concerns of Senators who feel that we must act now because the sensitive areas of their states are at risk. My concern...is that the acid rain control proposals which have been unveiled thus far are very expensive and will place thousands of jobs and hundreds of communities in our states at risk." **Acid Rain, 1983**, hearings before the Committee on Environment and Public Works, U.S. Senate [98-551] (Washington, D.C.: U.S.G.P.O., 1983), p. 365.

20. **Acid Precipitation and the Use of Fossil Fuels**, p. 158.

21. For an interesting discussion of the regional aspects of the politics of clean air, see Robert W. Crandall, **Controlling Industrial Pollution: The Economics and Politics of Clean Air** (Washington,

D.C.: Brookings Institution, 1983), pp. 110-130. See, two, also Robert W. Crandall, "An Acid Test for Congress?" **Regulation**, Vol. VIII (September/December 1984), pp. 21-28. See also Rochelle L. Stanfield, "Regional Tensions Complicate Search for an Acid Rain Remedy," **National Journal**, Vol. 16 (May 5, 1984), pp. 860-863; Michael S. McMahon and James M. Friedman, "A Midwestern Perspective on Acid Rain," **Federal Bar News and Journal**, Vol. 31 (February 1984), pp. 65, 69; Gordon L. Brady and Nancy A. Maloley, "Acid Rain: Science/Politics/Economics," **Ibid.**, pp. 59, 64.

22. See National Research Council, **Acid Deposition: Atmospheric Processes in Eastern North America** (Washington, D.C.: National Academy Press, 1983), pp. 116-126; Michael Oppenheimer, in **Acid Rain: A Technical Inquiry**, pp. 197-205; James N. Galloway, Gene E. Likens, Mark E. Hawley, "Acid Precipitation: Natural Versus Anthropogenic Components," **Science**, Vol. 226 (November 1984), pp. 289-831.

23. See, for example, G.E. Likens and C.V. Cogbill, "Acid Precipitation and the Northeastern United States," **Water Resources Research**, Vol. 10 (1974), pp. 1133-1137.

24. For a discussion of these issues, see D.A. Hansen and G.M. Hidy, "Review of Questions Regarding Rain Acidity Data," **Atmospheric Environment**, Vol. 16 (1982), pp. 2107-2126.

25. G.M. Hidy, D.A. Hansen, R.C. Henry, K. Ganeson, and J. Collings, "Trends in Historical Acid Precursor Emissions and Their Airborne and Precipitation Products," **APCA Journal**, Vol. 34 (April 1984), pp. 333-354. See also G.J. Stensland and R.G. Semonin, "Another Interpretation of the pH Trend in the United States," **Bulletin of the American Meteorological Society**, Vol. 63, No. 11 (November 1982), pp. 1277-1284.

26. D.A. Hansen and G.M. Hidy, "Review of Questions Regarding Acidity Data," op. cit., p. 2107.

27. See for example, "Are Utilities Obsolete?", **Business Week** (May 21, 1984), pp. 116-129. See also, Paul A. Carlson, "Electricity versus Gross National Product Growth--Past and Future," **Public Utilities Fortnightly**, Vol. 115 (January 10, 1985); Christopher Flavin, **Electricity's Future: The Shift to Efficiency and Small-Scale Power** (Washington, D.C.: Worldwatch Institute, November 1984); and Arthur A. Thompson, Jr., "New Driving Forces in the Electric Energy Marketplace--To a 'Death Spiral' or Vigorous Competition?", **Public Utilities Fortnightly**, Vol. 113 (June 21, 1984), pp. 31-40.

28. As one scientist working in this area has put it: "The key challenge...is to establish a source-receptor relationship as a prerequisite to the

development of a responsive control strategy to produce...a reduction in acidic deposition in sensitive areas. The source-receptor relationship must identify the magnitude of reductions in emissions of the substances contributing to or governing acid deposition and the geographic region in which such reductions must take place." Statement of Dr. Volker A. Mohnen, in **Acid Rain: A Technical Inquiry**, p. 4.

29. For an extensive technical discussion of these processes and the uncertainties involved, see U.S. Environmental Protection Agency, **The Acid Deposition Phenomenon and Its Effects: Critical Assessment Review Papers** (Washington, D.C.: U.S. EPA, Office of Research and Development, July 1984), Vol. I: "Atmospheric Sciences." (Hereinafter referred to as **Critical Assessment Documents**.) See section A-3 through section A-7.

30. Transmittal letter from National Research Council Chairman Frank Press, in **Acid Deposition: Atmospheric Processes in Eastern North America**, emphasis added. The qualifications included in the council's carefully worded statement are important in term of its applicability to policy. The council's conclusion clearly indicates that **all** states east of the Mississippi River must reduce all emissions **uniformly** (i.e., the same percentage reductions) in order for there to be a "corresponding fractional reduction in deposition" throughout the entire region. Moreover, the council noted that since its analysis is based on spatial distributions, "its applicability is limited to circumstances in which the spatial distribution of emissions is not changed." Because various analytic tools, such as atmospheric models, are not reliable, "we cannot objectively predict the consequences for deposition in ecologically sensitive areas of changing the spatial pattern of emissions in eastern North America, such as by reducing emissions in one area by a larger percentage than in other areas." **Ibid.**, p. 11.
There has been debate over the Council's conclusion, which was first aired in public before the Senate Committee on Environment and Public Works in late 1983. The policy significance of this debate is that "we are still unable to reliably predict deposition at specific sites or over sensitive regions of critical concern. This complicates designing optimized control strategies or providing strong assurances that a given reduction in emissions will result in a specified environmental benefit." Interagency Task Force on Acid Precipitation, **Annual Report, 1983** (Washington, D.C., 1983), p. 14.
For the debate over the council's statement on "linearity," see the testimony and discussion by

representatives of the National Research Council, Brookhaven National Laboratory, and the Environmental Defense Fund, in **Acid Rain, 1983**, pp. 381-410. See also, the documents submitted for the record, pp. 459-546, 555-634. See also **Implementation of the Acid Precipitation Act of 1980**, hearing before the Committee on Energy and Natural Resources, U.S. Senate (Washington, D.C.: U.S.G.P.O., April 30, 1984), pp. 1295-1394. Cf. Christian Seignour, Pradeep Saxena, Philip M. Roth, "Computer Simulations of the Atmospheric Chemistry of Sulfate and Nitrate Formation," **Science**, Vol. 225 (Sept. 7, 1984), pp. 1028-1030. On the basis of computer simulations of sulfate and nitrate formation in combination with reactive hydrocarbons, the authors suggest that SO_2 oxidation--involving hydrogen peroxide--may be oxident limited. This suggests that the chemistry of acid formation in the atmosphere may be "strongly nonlinear." See also J. Chamberlain, H. Foley, D. Hammer, G. MacDonald, O. Rothaus, M. Ruderman, **The Physics and Chemistry of Acid Precipitation** (SRI International, November 1981), esp. pp. V-28-V-29.
  31. Ibid., pp. 139-140.
  32. Ibid., p. 140.
  33. "Transport Processes," **Critical Assessment Documents**, Vol. I, pp. 3-90. The form of deposition (i.e., whether it is wet or dry) is an important factor. The National Research Council's 1983 report noted that "in much of the region of high ambient concentrations, dry deposition apparently dominates total deposition. This...suggests that dry deposition of SO_2 exceeds wet deposition in parts of the Ohio River Valley and the Ohio-Pennsylvania-western New York area but becomes progressively less important farther from the region of the major emissions, to the northeast or New England and Canada." **Atmospheric Deposition: Atmospheric Processes in North America**, p. 94.
  34. **Acid Deposition: Atmospheric Processes in Eastern North America**, p. 362.
  35. "Effects on Aquatic Chemistry," **Critical Assessment Documents**, Vol. II: "Effects Sciences," p. 4-1.
  36. **Ibid.**, pp. 1-166. See also Gary E. Glass, Thomas G. Brydges,and Orie L. Loucks (eds.) **Impact Assessment of Airborne Acidic Deposition on the Aquatic Environment of the United States and Canada** (Duluth, Minn.: Environmental Research Laboratory, U.S. Environmental Protection Agency Office of Research and Development, October 1981), pp. 128-133.
  37. For example, one recent study has suggested that an understanding of the contribution of long-range transport of acidic compounds "is impaired by the

paucity of quantitative data on the relative importance of inputs to surface waters 1) directly from the atmosphere, 2) indirectly from the atmosphere via runoff over or through the watershed, and 3) from the watershed due to internal generation." See M.D. Marcus, B.R. Parkhurst, and F.E. Payne, **An Assessment of the Relationship Among Acidifying Deposition, Surface Water Acidification, and Fish Populations in North America** (Palo Alto, Calif.: Electric Power Research Institute, June 1983), Vol. I, p. 3-1.

38. The term "acidification" when applied to lakes and streams refers to a decrease in alkalinity over a period of time, or as "the loss of alkalinity." Alkalinity refers to the acid-neutralizing capacity of a body of water. See "Effects on Aquatic Chemistry," **Critical Assessment Documents**, Vol. II, pp. 4-5-4-6.

39. **Ibid.**, p. 4-35. For an empirical analysis of these factors in relation to three lakes in the Adirondacks, see **The Integrated Lake-Watershed Acidification Study** (Palo Alto, Calif.: Electric Power Research Institute, November 1984), four volumes.

40. See, **The Integrated Lake-Watershed Acidification Study, Vol. 4: "Summary of Major Results."**

41. Critical Assessment Documents, Vo. II, p. 4-25. See also, Glass, et.al., **Impact Assessment of Airborne Acidic Deposition on the Aquatic Environment of the United States and Canada**, p. 2. See, also, the statement by Orie L. Loucks, in **Acid Rain: A Technical Inquiry**, pp. 399-416; Statement of Jerald L. Schnoor, in **Implementation of the Acid Precipitation Act of 1980**, pp. 411-420; statement of Gary Glass, in **Acid Rain, 1983**, pp. 105-144; and Gail Einbender, Allan Bakalian, Thomas Hall, Peter Hoagland, and Kenneth S. Kamlet, "The Case for Immediate Controls on Acid Rain," in **Acid Precipitation and the Use of Fossil Fuels**, pp. 1322-1324.

42. See, for example, the statement of Dennis R. Lundervill, Director of the New Hampshire Air Resources Angency, in **Effects of Acid Rain**, hearings before the Subcommittee on Energy Conservation and Supply of the Committee on Energy and Natural Resources. U.S. Senate [Publication No. 96-126] (Washington, D.C.: U.S.G.P.O., June 21, 1980), pp. 53-55. An extensive review of the scientific literature analyzing acidification trend data in aquatic systems in different states and Canada is in **Critical Assessment Documents**, Vol. II, pp. 4-64-4-113.

43. Interagency Task Force on Acid Precipitation, Annual Report, 1983, p. 41.

44. Norman E. Peters, Roy A. Schroeder, and David E. Troutman, **Temporal Trends in the Acidity of Precipitation and Surface Waters of New York**, United

States Geological Survey Water Supply Paper 2188 (Washington, D.C.: U.S.G.P.O., 1982), p. 32. In a later report U.S.G.S. found that in the "northeastern quarter of the country SO_2 emissions have decreased over the past 15 years and the trends in the...chemical characteristics of...streams are consistent with a hypothesis of decreased acid deposition in that region." Richard A. Smith and Richard B. Alexander, **Evidence for Acid Precipitation-Induced Trends in Stream Chemistry at Hydrologic Bench-Mark Stations,** Geological Survey Circular 910 (Washington, D.C.: U.S.G.P.O., 1983), p. 12.

45. See National Research Council, **Acid Deposition: Atmospheric Processes in Eastern North America,** p. 119.

46. For a brief discussion of this issue, see **Critical Assessment Documents,** Vol. II, pp. 4-39-4-45.

47. "The available information well documents a trend in surface water acidification attributable to precipitation acidification in the Northeastern United States and Southeastern Canada. The Canadian sites mostly experienced acidification due to local sources. In the Northeastern United States, regional changes in precipitation chemistry apparently acidified lakes and streams in new England, New York, and New Jersey. The acidification appears to have occurred before the mid- to late 1960's, because data collected continously since that time shows a lack of further acidification in New York streams, and a slight recovery of some acidified streams in scattered locations throughout the Northeast. The findings of these stream chemistry studies are consistent with a region-wide decrease in SO_2 emissions in the Northeast for the period 1965-1980...In the West, most reports of the acidification of surface water have reflected local conditions...The source of acidity is estimated to be local urban sources." John T. Turk, **An Evaluation of Trends in the Acidity of Precipitation and the Related Acidification of Surface Water in North America,** U.S. Geological Survey Water-Supply paper 2249 (Washington, D.C.: U.S.G.P.O., 1983), p. 14.

Evidence of the acidification of surface waters in the West complicates the political situation immensely, since representatives of western states will not support a national acid rain program. As Senator Alan K. Simpson (R-Wyo.) has argued, "acid rain is not recognized as a widespread problem in the West. In fact, only a few 'hot spots' exist where acid precipitation is causing limited environmental damage...We should be wary of over reacting to the acid rain situation by requiring that all 50 states implement massive sulfur reductions." Statement of Senator Alan K. Simpson, in **Acid Rain, 1983.** p. 362.

Some media attention has been given to the growing evidence that acid rain is occurring in the West, and that the environmental damage is not limited to a few "hot spots." See, for example, Philip Shabecoff, "Acid Rain Attacks Environment Beyond the Northeast," **New York Times** (January 29, 1985), p. C1; and Michael Oppenheimer, "Acid Rain has Spread to the West," **Los Angeles Times**, August 28, 1983, p. D3. Environmental organizations have also begun to focus more attention on the issue of acid rain in the West. See Michael Oppenheimer and Robert E. Yuhnke, "Acid Deposition, Smelter Emissions, and the Linearity Issue in the Western United States." (Environmental Defense Fund, unpublished paper, n.d.); **'Acid Rain' in the Intermountain West.** [A Report by the Environmental Defense Fund of a Survey of Acid Rain Research in the Rocky Mountain Region.] (January 1982). See also, "Acid Deposition in Colorado," prepared by the Colorado Department of Health, Air Pollution Control Division (October 1983).

48. Statement of William D. Ruckelshaus, Administrator of the U.S. Environmental Protection Agency before the Senate Committee on environment and Public Works (February 2, 1984.) Thus, even if acid deposition levels have stabilized, according to this hypothesis, there will be a point at which a lake or stream will suddenly become totally acidified. As one scientist has expressed this hypothesis, "acid rain is a cumulative phenomenon. The fact that the amount of acid rain we are adding each year hasn't necessarily changed dramatically doesn't mean you won't reach a dramatic endpoint." Dr. Roy Gould, in **Acid Rain: A Technical Inquiry**, p.66. A discussion of these two hypotheses appears in **Acid Deposition: Processes of lake Acidification** (Washington, D.C.: National Academy Press, 1984). A general description of the acidification process is found on pp. 2-3.

49. **Acid Deposition: Processes of Lake Acidification**, p. 6.

50. Ruckelshaus, op. cit.

51. **Acid Deposition: Processes of Lake Acidification**, p. 11.

52. See for example, Peter J. Dillon, Norman D. Yan, Harold Harvey, "Acidic Deposition: Effects on Aquatic Ecosystems," in **CRC Critical Reviews in Environmental Control**, Vol. 13, No. 3, pp. 176-194; Robert A. Boyle, "A Rain of Death on the Striper?" **Sports Illustrated**, Vol. 60 (April 23, 1984), pp. 40-42, 44, 51, 54; and Angus Phillips, "Acid Rain Strikes Close to Home, With Yellow Perch the Unfortunate Witness," **Washington Post**, March 24, 1985, p. G6.

53. See **Critical Assessment Documents**, Vol. II, p. 5-126. See also, pp. 5-125 - 5-128.

54. **Ibid.**, p. 5-159. However, the evidence linking changes in aquatic communities and acidification should be treated with caution. For example, most of the information on the effects of low pH on fish and other biota is based on laboratory experiments. Such studies "have yielded data on only a few species. Predicting community-level changes from laboratory bioassays on a few species is difficult. A species may experience reduced growth or reproduction in the laboratory at a low pH, but may prosper in an acidified lake at the same pH if its competitors suffer even greater reduction in growth and reproduction." **Ibid.** See pp. 5-1, 5-2. Furthermore, "with few exceptions, the mechanism of acidification effects (whatever the cause) on fish, amphibians, and other aquatic organisms are poorly understood. Biological changes can occur as pH declines below 6.0 and can become increasingly severe as pH declines further...These effects are currently known only qualitatively, except for the dose-response relationship between aluminum and selected fish species." Interagency Task Force on Acid Precipitation, **Annual Report 1983**, p. 42. See also **Critical Assessment Documents**, Vol. II, pp. 5-128, 5-126. For an analysis of the relationship between acid deposition and the mobilization of metals such as aluminum in aquatic and terrestrial systems, see **Acid Deposition: Effects on Geochemical Cycling and Biological Availability of Trace Elements** [Subgroup on Metals of the Tri-Academy Committee on Acid Deposition] (Washington, D.C.: National Academy Press, 1985).

55. For a review of studies which suggest a link between air pollution generally and damage to tree species in the U.S. and Europe, see George H. Tomlinson, "Air Pollution and Forest Decline," **Environmental Science and Technology**, Vo. 17 (1983), pp. 246A-256A. See also, Richard Plochmann, "Air Pollution and the Dying Forests of Europe," **American Forests** June 1984, in **Effects of Air Pollution and Acid Rain on Forest Decline** oversight hearing before the Subcommittee on Mining, Forest Management, and Bonneville Power Administration of the Committee on Interior and Insular Affairs, House of Representatives [Serial 98-30] (Washington, D.C.: U.S.G.P.O., 1984), pp. 84-88. "Among scientists from different schools there is a widespread agreement that the recent forest damages are predominantly to be ascribed to anthropogenic air pollutants. Without anthropogenic air pollution there would not be forest damages to the present extent. There is an animated scientific discussion only on the questions of which pollutants are active to what degree on which sites and what are the response to mechanisms for triggering the damages,

and how the contributing natural stress factors should be weighed." Florian Scholz, "Report on Effects of Acidifying and Other Air Pollutants on Forests," in **ibid.**, p. 113.

56. Statement of Baron Franz Reiderer Von Paar, in **Effects of Air Pollution and Acid Rain on Forest Decline**, pp. 20-21. It is important to note that there are substantial differences between the conditions apparently affecting the forests of Germany and the situation in the United States. These differences may limit the applicability of the West German situation to the situation in this country. For example, levels of deposition are significantly higher in West Germany than in the United States. Population density is higher in Germany, and the composition of American forests is significantly different. However, the German experience does indicate that serious terrestrial effects may be linked to high levels of air pollution.

57. Richard Plochmann, "Air Pollution and the Dying Forests of Europe," op. cit., p. 87. See also, **Acid Rain in Europe**, pp. 2-8.

58. Plochmann, op.cit., p. 85. See also Peter Schutt and Ellis B. Cowling, "Waldsterben--A General Decline of Forests in Central Europe: Symptoms, Development, and Possible Causes of a Beginning Breakdown of Forest Ecosystems." (Manuscript submitted to **Plant Disease**).

59. Arthur H. Johnson and Thomas G. Siccama, "Acid Deposition and Fost Decline," **Environmental Science and Technology**, Vol. 17 (1983), pp. 296A-297A. See also, Arthur H. Johnson, Andrew Friedland and Thomas C. Siccama, "Recent Changes in the Growth of Forest Trees in the Northeastern United States," in **Air Pollution and the Productivity of the Forest**, pp. 121-142; and **Critical Assessment Documents**, Vol. II, pp. 3-35 - 3-38.

60. Ellis B. Cowling, "Conclusions Regarding the Decline of Forests in North America and Central Europe" (unpublished paper, Fall 1984), pp. 1-2.

61. **Ibid.**, pp. 2-3. See also Johnson and Siccama, "Acid Deposition and Forest Decline," op. cit., p. 304A; and **Critical Assessment Documents**, Vol. II, pp. 3-17 - 3-26.

62. Schutt and Cowling, "Waldsterben--A General Decline of Forests in Central Europe," p. 26. For a critique of this hypothesis, see **Critical Assessment Documents**, Vol. II, p. 3-38. See the discussion of this hypothesis relative to data developed for red spruce in the northeastern United States in Johnson, Friedland, and Siccama, "Recent Changes in the Growth of Forest Trees in the Northeastern United States," op. cit., pp. 134-137. Cf., the statement of Nye M.

johnson, which suggests that this hypothesis has considerable merit, in **Acid Rain: A Technical Inquiry,** pp. 230-231.

63. **Critical Assessment Documents,** Vol. II, p. 3-3. See also Johnson and Siccama, "Acid Depositon and Forest Decline," op. cit., p. 300A, for a brief critique of this hypothesis. However, a report of the Society of American Foresters suggests: "The most important regional pollutants, in addition to acid deposition, include ozone and trace metals. Wide areas of the temperate forest in North America are subject to concurrent exposure to these three contaminants. Only in the case of ozone are clear direct forest tree responses evident." **Report of the SAF Task Force on the Effects of Acidic Deposition on Forest Ecosystems.** (Bethesda, Md.: Society of American Foresters, May 1, 1984), p. 32.

64. Schutt and Cowling, op. cit., p. 27.

65. **Critical Assessment Documents,** Vol. II, p. 3-40. See also, pp. 3-9 - 3-13.

66. Schutt and Cowling, op. cit., p. 28. See also, Johnson and Siccama, op. cit., p. 304A. A variant of this hypothesis has been suggested with respect to red spruce dieback. "The spruce decline appears to be a stress-related disease. The trees are probably predisposed to decline by the site conditions whereby some short-term stress, possibly the drought of the early 1960's, triggered a loss of vigor, and where biotic stress imposed by fungal attack is sufficient to cause widespread mortality. Acidic deposition could act to intensify the predisposing stresses, exacerbate the effects of the triggering streams or increase the susceptibility to fungal attack..." **Critical Assessment Documents,** Vol. II, p. 3-39.

67. Schutt and Cowling, p. 29. See, too, Society of American Foresters, op. cit., pp. 16-18. It is interesting to note, however, that NO_x or related compounds, such as ozone, seem to figure prominently in many of the hypotheses regarding forest decline. This suggests that policy efforts to address the acid rain problem which emphasize SO_2 reductions may not address the problem of forest decline adequately.

68. Concern has been expressed that acid deposition may have an adverse impact on soil, fertility, the availability of nutrients, plant vitality and water quality. The acidification of soils has apparently become a serious new concern in Europe. European scientists have "expressed a fear that soil acidification could irreversibly alter the natural ecosystems, with significant impacts on forest and agricultural productivity, and groundwater quality." **Acid Rain in Europe,** p. 9. In the United States, "the long-term effect (i.e., over decades or centuries) of

acidic deposition can be expected to remove cations from forest soils, but it is not clear whether this will reduce available cations and enhance acidification of soils." **Critical Assessment Documents,** Vol. II, p. 2-55. Increased aluminum mobility in soils as a result of acid deposition is a serious concern. "The increased mobilization of Al in uncultivated, acid soils is probably the most significant effect of acidic deposition on soils as they influence terrestrial plant growth and aquatic systems," **Ibid.** However, it is "unlikely that many soils will be significantly acidified by acid rain at current input levels in the United States," **Ibid.**, p. 2-33. The effects of acid deposition on public monuments and buildings has been documented. However, damage to materials "is believed to be due more to acid deposition from local sources than to acidified rain produced from long-range transport of pollutants and their reaction products," **Ibid.**, p. 7-54.

Part I

The Political Context

The political stalemate over acid rain policy in the United States has been a subject of widespread commentary. At first glance, given the degree of complexity and uncertainty surrounding the science of acid rain and the magnitude of the economic and other costs associated with mitigation of this environmental problem, it may not seem startling that a deadlock--with no resolution in sight-- has taken hold in the legislative policy process. Yet Congress produced about a dozen major environmental statutes from 1969 to 1980; and these acts included complex legislation addressing a variety of environmental issues like air and water quality, the disposal of toxic wastes, and the control of pesticides and other toxic substances. For each of these issues many, if not all, of the same general political characteristics as are found in the acid rain debate were manifest. For example, regional considerations played a significant role in the dispute over the Clean Air Act in 1970, and again in 1977 when important amendments to that act were added. And, when the Toxic Substances Control Act was being considered in the Congress a few years ago, the science bearing on these substances was no more certain and complete.

The two chapters in this part explore different facets of the bases of the acid rain policy impasse. Written as "an exercise in critical policy analysis," the chapter by Ernest J. Yanarella examines the "political ideological dimensions of the policy framework" grounding many of the essential terms of the debate. The chapter by Phillip Roeder and Timothy Johnson analyzes "public opinion on the problem of acid rain"--another important, but more evanescent, aspect of the political context of the controversy.

In general, Yanarella's essay illuminates the broad institutional features and fundamental political-cultural dynamics of the American political system, and shows how they contribute to, and even exacerbate, the current policy stalemate. Adopting a critical posture toward the literature on political immobilism in advanced capitalist societies in the West, it analyses the complex interplay between and among such factors as the operation of "interest-group liberalism" in the Congressional arena, the historically weak administrative apparatus of the State, the rise of administrative centralization in the presidency, the dominance of corporate interests and their over-representation at key access points in the political process, and the contribution of America's two-party system to the "regional polarization" which has been such a visible feature of the acid rain debate.

Yanarella also explores the "bureaucratization of environmental protection" in this country, a consequence of the creation of the Environmental Protection Agency (EPA) during the Nixon administration. He argues that the institutionalization and professionalization of environmental lobbies elevated "environmentalism" to the status of a "legitimate political interest," but also meant that genuine ecological priorities came to be pursued in terms of interest-group politics, the operation of which defused the latent, though limited, creative potential of the environmental movement. Thus, Yanarella suggests, the potential for fundamental ecological reconstruction in the American polity has been largely eclipsed by environmental reformism.

In the context of the political economy of the United States, treated by Yanarella as an advanced capitalist (and increasingly technocratic) society, he discerns an "irreconcilability" between an acid rain control policy grounded in a sense of ecological urgency and the "interest group syndicalism of American politics." This is not to suggest that the current impasse cannot be overcome. For he suggests that it may be possible to end the stalemate with the "right" (i.e., politically feasible) mix of "technological fixes" and "regulatory palliatives." In that event, however, the acid rain debate would be resolved in narrowly "reformist" terms without confronting the broader political-ecological requirements of dealing with industrial pollution.

Roeder and Johnson's survey of public opinion toward acid rain draws upon the vast social science literature on the structure and attributes of American political attitudes. Using polling data from Gallup, Harris, Roper, and a variety of statewide surveys, these social researchers examine the level of public concern, public knowledge about the causes and environmental effects of acid rain, and public attitudes about paying the costs of addressing this environmental problem. Their excavation of the data leads Roeder and Johnson to conclude that public awareness of acid rain "has increased dramatically over a brief period of time." On the other hand, they note that the issue is not particularly salient when compared to political attitudes toward other issues of public concern. It would appear that other issues which are perceived as involving threats to the environment and public health (e.g., toxic chemicals, hazardous waste, drinking water) are of greater concern to Americans than the dangers of acid rain.

Roeder and Johnson find that, in general, the public is not well informed about the causes and consequences of acid rain, although it does appear to be "reasonably informed" about proposals for action to grapple with the problem of acid rain and about the costs of such proposals. But, when a "personal price-tag is attached to the action," there is a declining level of enthusiasm--especially as costs to individuals approach and exceed amounts of $100 per year. In concluding that public attitudes on acid rain are weakly-held and subject to significant shifts, Roeder and Johnson imply that the structure of public opinion on this issue parallels the terms of the national debate. To some extent, this coincidence may reflect the results of efforts by both sides to influence public perceptions of the acid rain issue. This view seems consistent with Yanarella's suggestion that neither proponents nor opponents of acid rain legislation will be able to mobilize public opinion sufficiently to exercise decisive leverage in the political contest.

2
The Foundations of Policy Immobilism over Acid Rain Control

Ernest J. Yanarella

INTRODUCTION

For more than five years, the political controversy over acid precipitation has raged in the United States. Not long ago, the acid rain problem was characterized by EPA Administrator William Ruckelshaus as "the most difficult, complex public policy issue" he ever faced and probably one of the "cosmic issues" of our time that this nation and the rest of the world might be forced to address.[1] Yet despite countless technical studies, numerous policy recommendations, and dozens of legislative bills generated, resolution of this knotty problem remains elusive. Competing scientific, economic, and political interests and perspectives have been injected into the struggle and have influenced the changing foci of the public debate over the nature, impacts, and costs of acid rain and its mitigation. While the precise contours of an acid rain control strategy that may eventually be instituted in the United States are unclear, critical policy analysis reveals the major sources of blockage in the American policy process.

As an exercise in critical policy analysis, this essay views the phenomenon of acid deposition in its multiple forms as a significant, but uncertain, environmental and public health problem enmeshed in a web of other energy, environmental, and social issues that must be confronted. This chapter illuminates the foundations of policy immobilism by delving into the political and ideological dimensions of the policy framework within which this issue is being contested. This standpoint is independent of the scientific agnosticism of coal industry supporters and the doomsaying of professional environmentalist organizations--one informed by the perspective of political ecology.

THE MAKING OF A POLICY STALEMATE

Although concern for the problem of acid rain can be traced as far back as 1852 when the English chemist Robert Angus Smith investigated the atmospheric chemistry of what he would eventually term "acid rain" falling in and around the city of Manchester, England, it was not until the early 1970s that serious scientific work and intermittent political interest occurred in the United States [2]. Since the mid-seventies, however, mounting political pressures have been generated and an impressive coalition of social and economic interest groups has emerged on one side or another into the "acid rain wars" of the late seventies and early eighties.

Between 1976 and 1980, early political skirmishing over the acid rain issue took place at the national and trans-national levels. Domestically, the Carter administration showed particular interest in grappling with the acid rain problem even though scientific research into the many dimensions of the problem in the United States was skimpy and inflation and other economic woes inhibited federal initiative to fashion a comprehensive and well-funded mitigation policy. The political outcome of this presidential interest and early lobbying struggle was the National Acid Precipitation Act of 1980-- a comprehensive ten-year research program administered by an interagency task force to explore the sources and impacts of acid deposition.[3] Meanwhile, Canadian worries over the mounting evidence of transboundary acid rain pollution precipitated bi-lateral negotiations and technical workshops to work toward a formal U.S.-Canadian agreement for controlling transboundary pollution.[4]

In the period between this initial phase of the "acid rain wars" and the present, the politics of acid rain intensified and the early resolution of the controversy was challenged by the right and on the left. The Reagan administration decided to stall diplomatic negotiations with Canada on a bi-lateral treaty that was to alleviate transfrontier acid rain damage. Reversing the steady progress toward a common understanding of the nature of the acid rain problem and a joint intention to mitigate the consequences of each country's transboundary pollution, the administration challenged the meaning of the Memorandum of Intent signed by each country in 1979, questioned the findings and conclusions of the Bilateral Research Consultation Group, and replaced scientists and diplomats with others who effectively stalled the talks and brought the technical working groups and diplomatic bargaining back to square one.[5]

Meanwhile, congressional deliberations over the reauthorization of the Clean Air Act of 1970 provided a new forum within which strategies for mitigating acid rain damage in the United States could be debated. Having lost in their efforts to compel installation of "scrubbers" on old coal-fired plants, various environmentalist groups seized this opportunity to renew this quest in the face of mounting evidence that the favored technological fix--tall smokestacks--had only exacerbated the problem of regional pollution downstream. Despite administration attempts to dilute portions of the Act in order to relieve industry of some of its requirements, the environmentalist debate has crowded out other issues and made acid precipitation control the preeminent one.[6] As a consequence of these international and domestic developments, a policy impasse materialized as public debate divided into two principled and apparently irreconcilable coalitions of power. Since its inception, neither the pro-control alliance nor its anti-control counterpart has been able to muster sufficient political clout to win the day in the renewed acid rain dispute.

For nearly two years, federal acid rain policy was immobilized by this stand-off. The two competing constellations of power crystallized into a recognizable set of constituents (see Figure 2.1), while influencing executive perceptions, lobbying Congress, and wooing public opinion. For the anti-control alliance, opposition to investment into any control technologies led it to fashion a pro-industry strategy which included the following features:

   --Postpone mitigation efforts: defer mitigation programs involving sizeable funding because some control technologies and programs would jeopardize key interests;

   --Buy time: stress the non-crisis nature of the environmental problem and the lack of an adequate scientific basis for prudent policy response in order to gain latitude for mobilizing political resources to shape or block any policies touching on major corporate interests;

   --Tout an alternative policy of long-term research programs: propose government-sponsored and corporate-funded research projects lasting up to a decade or, ostensibly, to obtain a fuller picture of acid rain causes and consequences;

   --Manipulate the methods and values of science: exploit the inherent caution, conservatism, and skepticism of natural scientists faced with an uncertain environmental phenomenon;

   --If all else fails, make sure the consumer/citizen pays: in the event that a control policy seems politically inevitable, seek to shift the burden of costs to consumers or citizens of the nation-at-large rather than to the polluters.

On the other side, the pro-control alliance--motivated by its goal to see abatement technologies and other mitigation programs put in place as quickly as possible to alleviate the consequences of acid deposition--designed an environmental strategy composed

FIGURE 2.1
Competing political alliances

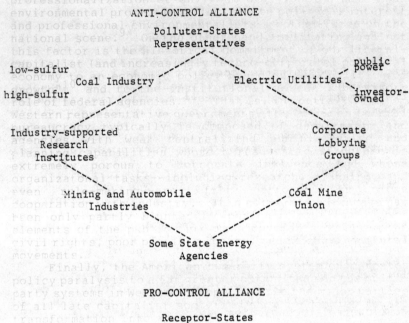

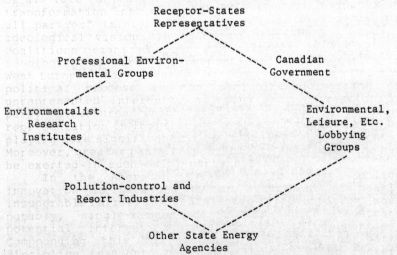

of the following aspects:

--**Generate a crisis atmosphere around the acid rain issue:** portray the threat in emotionally-charged, even apocalyptic, terms ("death from the sky," "rain of terror," "poison from above," "deadly rain," and "dropping acid") posing dangers which necessitate urgent, massive efforts;

--**Accentuate the negative:** highlight those parts of all scientific studies and technical reports supporting the crisis nature of the policy problem, drawing upon studies from the life sciences (forestry, fish, plant, animal, humans) in particular to demonstrate the breadth of impact of acid rain;

--**Exploit the mass media:** skillfully use the potential of the mass media to educate and manipulate public attitudes and to underscore the crisis nature of the acid rain situation;

--**Build an environmentalist/conservationist coalition:** draw upon traditional environmentalist constituencies for the purpose of generating funds, issue visibility, and political muscle for the congressional interest-group battle.

With these two contending strategies and coalitions in mind, one might ask: Why did the politics of acid rain tend to resolve itself into a policy stalemate? And, what were the factors tending to break up the policy log-jam for the first eight to ten months of 1983?

As a number of political analysts have observed, policy immobilism appears to be endemic to most advanced capitalist societies.[7] One key feature of policy immobilism is the apparent contradictory situation of increased state activity coupled with decreased policy choices. On the West European scene, the tendency of late capitalist polities to succumb to this malady has been variously attributed to: the tensions and contradictions of system legitimation and capital accumulation thrust upon the state; the growth of administrative bureaucracy and its absorption of political tasks which it cannot fulfill; and the depoliticization of parties and the consequent dilution of their historical role of mediating and aggregating citizen interests and grievances.[8]

On the American scene where the influence of interest-group liberalism has been pervasive at the middle levels of power, the nature of broker politics in the legislative domain tends to prompt congressional representatives simply to broker the established interests of the economically privileged and politically well-organized. And, while broker politics is less visible in the executive realm, where dominant corporate interests tend to hold sway, one can usually count on brokering to take place when re-election time nears in an effort to forge a winning coalition. Secondly, despite the greater financial base and easier political access of corporate interests in our political system, the institutionalization of environmental protection within the federal bureaucracy in 1969 and the accompanying professionalization of environmental lobbies have made environmental protection a legitimate political interest and professional environmentalists a potent force on the national scene.[9] On the other hand, militating against this factor is the historical commitment of our liberal-capitalist (and increasingly techno-corporate) political economy to an economic policy based on the "politics of growth"[10] and to the institutionally weak supervisory role of federal agencies.[11] That is, in contrast to other Western representative governments, the American federal bureaucracy typically is composed of departments and agencies with weak centralizing administrative and planning capabilities whose bureaucratic structure is extremely porous to corporate influence and whose organizatonal tasks--including research, planning, and even rule-making--are often performed in close cooperation with industry. This corporate advantage has been only partly blunted by periodic mobilization of elements of the public into mass movements, such as the civil rights, poor people's, peace, and now environmental movements.

Finally, the American two-party system exacerbates policy paralysis to a far greater extent than do the multi-party systems in Western Europe. While the major parties of all late capitalist societies show tendencies toward transformation into what Otto Kirchheimer calls "catch-all parties" (i.e., parties which over time sacrifice ideological visions in order to constitute governing coalitions organized around the political center of the ideological spectrum),[12] proportional representation in West European party systems has frequently opened up the political process to the periodic aggregation of unrepresented interests, allowing them expression in party structures which stand a chance of gaining representation in legislative bodies and perhaps even of playing more significant roles in coalition governments. Moreover, greater party responsibility and discipline can be exerted in such multi-party systems.

In the American system by contrast, policy innovation is weakened by an assortment of nearly insuperable barriers to legislative representation (most notably, single-member districts) which frustrate potential interest aggregation by third parties. Compounding this circumstance, the lack of party discipline fragments party loyalty and often fosters regional coordination of power among politicians who--regardless of party label--tend to press for policies supportive of the dominant economic interests spanning their region. Illustrative of the influence of these factors is the way the impasse over acid rain policy has been effected by the bitter regional conflict among political forces in the Northeast, the Midwest, the South, and the West--pitting governors, state agencies, and leading industries in one area against their rivals in the other.[13] Indeed, as various observers pointed out and some political participants concede, the diverse bills presenting proposals for alleviating the effects of acid deposition reflect the specific regional interests of their sponsors. This polarization has also been heightened by the regional representation of members on the House and Senate committees which are reviewing these legislative proposals.[14]

In addition, the policy stalemate has been fundamentally shaped by the split within the scientific community over issues relating to the causes, extent, and dangers of acid rain to the environment and to public health. For example, one need only compare the conflicting scientific assessments and contradictory conclusions on these issues reached by Dr. Michael Oppenheimer and Dr. Volker Mohnen, two noted atmospheric scientists, to get a sense of how confusing the scientific picture presents itself to politicians and the public.[15] This fragmentation has assured both sides in the dispute of ample scientific support for their opposing definitions of the situation and favored strategies of action or inaction while simultaneously eroding public confidence in the capacity of science to offer valid and unambiguous answers to this knotty socio-technical issue. Yet, as many scientists on both sides of the issue have readily admitted, scientific understanding of the acid rain phenomenon will always be incomplete, so "absolute proof in the legal or even the scientific sense is unlikely ever to be obtained."[16] In this regard, the call by politicians for scientific resolution of this issue is really an effort to evade their responsibility for rendering a political judgment on this issue.

Certainly, by the first few months of this year, the frozen state of this political controversy over acid rain began to show distinct signs of thawing.[17] Evident movement toward a transformation in the policymaking context was, first of all, aided by the Reagan administration's overstepping of the bounds of public consensus on environmental policy caused in part by its gross misreading of the perceived electoral mandate given the President in 1980 to relieve industry of "undue regulatory burden by government." The policy reversals of the Burford-directed Environmental Protection Agency--including the nightmares and horrors triggered by an administrative agency packed with industry representatives--can be traced to a basic misconstrual of public sentiment over environmental issues and to a woeful ignorance of the existence and durability of environmental consciousness among a clear majority of the mass public.[18] Then, too, the release that year of scientifically cautious, but politically devastating findings and conclusions in major reports by scientific panels of irreproachable credentials was a second factor altering the framework of policy-making on this issue. The combined weight of the Interagency Task Force on Acid Precipitation second annual report, the White House's OSTP interim report, and the National Academy of Sciences study effectively undercut key arguments and contentions upon which the pro-industry/anti-control case was grounded.[19] Magnifying the political impact of these studies, the mass media and environmental spokespeople depicted in unequivocal, blanket terms the often highly qualified, guarded wording of those reports.

Thirdly, certain political and scientific developments in multi-national conferences on the acidification of the environment led to the growing isolation of the formal American position from the inchoate consensus around this troubling issue among European nations.[20] Thus, the emerging recognition of the world-wide threat posed by acid deposition meant that such developments in Western Europe and Canada would highlight the problem and produce tangible political resources for environmentalist/pro-control forces in the domestic political arena. Even more politically damaging was the unravelling of the opposition coalition caused by the concern of several constituents of the anti-control alliance (e.g., the low-sulfur coal industry, the automobile industry, and the public power utilities) that their narrower economic interests be protected in the face of the increasing likelihood of some administrative policy initiative or legislative control policy percolating out of the changed political environment surrounding the issue.[21]

A more diffuse and evanescent, yet in my opinion no less controlling, factor moving the acid rain controversy beyond policy immobilism has been the trickling up from segments of the public of an altered attitude toward the broad spectrum of environmental hazards and their effects upon future generations. One palpable effect of Three Mile Island, Love Canal, and Times Beach has been increasing public skepticism of the established levels of acceptable risk to human beings and the environment from radiation, chemicals, and environmental pollutants. Growing numbers have come to the view that an involuntary risk or danger to future generations imposed by others (like the EPA and the NRC) ought to be signficantly lower than self-imposed risks. In this respect, this rapidly evolving attitude suggests, in Barry Commoner's words, that "the public has now become aware that environmental pollutants [including, I would argue, acid deposition] represent an assault by the present generation not merely on involuntary living victims who have some recourse, however difficult--but on generations not yet born, and therefore utterly defenseless."[22]

The significance of this shift in public perception and attitude toward such pollutants is two-fold. First, it destroys the whole foundation of the risk/benefit calculus that some scientists and bureaucrats have tried to sell the American public in considering, for example, the comparative risks and benefits of acid rain mitigation. Thus, David Stockman's recent utilitarian arguments fall upon increasingly deaf public ears when he says: "I kept reading these stories that there are a hundred seventy dead lakes in New York that will no longer carry and fish or aquatic wildlife, and it occurred to me to ask the question...How much are the fish worth in these hundred seventy lakes that account for four percent of the total lake area of New York? Does it make any sense to spend billions of dollars controlling emissions from sources in Ohio and elsewhere if you're talking about **very marginal volumes of dollar values**, either in recreational or commercial terms?"[23] Secondly, apart from the influence of the environmental lobby in Washington, this altered public attitude and its impact upon the operating political climate of opinion explains why congressional representatives have perceived the existence of an aroused public demanding a control policy when there has not been a single national public opinion poll conducted surveying mass attitudes on the acid rain issue.

THE REMAKING OF A POLICY STALEMATE

Certainly by late summer of 1983, political resolution of the paralysis over legislating means of alleviating the environmental repercussions of acidic deposition seemed imminent. In response to the host of factors enumerated above, Reagan appointed William Ruckelshaus once again to head EPA in an obvious move to shore up sagging public confidence in the administration's commitment to environmental protection. And, at the public announcement of this personnel change, Reagan charged the new administrator with the task of grappling with the acid rain problem as a top priority of the federal environmental protection agency.[24] Having miscontrued his electoral madate, Reagan was evidently moving to repair the political damage stemming from his policy of populating regulatory agencies (like EPA) at the beginning of his first term with representatives or supporters of the industries whose actions those governmental organizations were supposed to monitor. With the ungluing of the anti-control coalition and the undermining of the key rationale for postponement, the political ground for policy compromise seemed prepared.
The emerging interim resolution of the political problem within EPA took the form of a limited, experimental program focused upon the environmentally-damaged and politically sensitive Northeast. According to reports, it was to be a modest $1.5-2.5 billion program

targeted at relieving the acid rain problem in the Adirondacks region of New York and New England.[25] As it unfolded, its goal was to limit sulfur emissions by a total of 4-5 million tons annually, and its key requirements were to include reduction of sulfur emissions by approximately 50% in Ohio and West Virginia, by about 30% in Pennsylvania, and by something like 15% in New York, with smaller reductions in New Hampshire and Vermont. Moreover, rather than legislating by federal mandate any specific method to achieve reduced sulfur levels, it was tentatively decided to permit the states involved to choose whatever method they preferred to attain those limits. Left unclear was how the program would be funded--although there was some interest within EPA in a national tax attached to consumer electric bills. From the institutional perspective of EPA, this modest, flexible program offered "something to everyone" and had the further benefit of being capable of being expanded to other regions and eventually to the entire nation as political conditions permitted and as scientific findings warranted.

When the plan was brought before the Cabinet Council on Natural Resources and the Environment in October 1983, it was extensively criticized and bitterly opposed. Why was it rejected and how was a control policy stalemate restored? In the first place, EPA apparently could not overcome the antagonism and power of the intra-administration opposition at the elite level within the Executive Branch.[26] Resistance by the Office of Management and Budget (OMB) and the Department of Energy (DOE) to even a modest test program on the grounds that it was too costly and premature demonstrated that the administration's inner elite coterie remained insulated from the changing political perceptions of broader elements operating within the middle levels of power in the American system. Although the political currency of votes within the Congress and among the electorate were influential, the pressures of the mounting budgetary deficit and the pseudo-free market mentality informing key Reagan appointees produced an unfavorable cost-benefit calculation regarding this control option.

To elaborate, OMB director David Stockman's politically-veiled ideological opposition to acid rain control legislation is well-known and has been longstanding. Thus the politicized role and policy stance taken by the director of OMB on this issue are hardly surprising. However, what is less widely recognized is that the politicization of OMB in the policymaking process stems not so much from the selection of Stockman as its director as from its legislative history and institutional origins in the Nixon administration. Alan Wolfe has noted how, as the old Bureau of the Budget assumed more and more crucial functions within the federal government, pressures grew to restructure it in a way which compromised its administrative competency and neutrality and turned it into a political arm of the governing political regime.[27]

The opposition of the Department of Energy, on the other hand, finds its impetus in the market-steered and growth-oriented policies built into this weak energy management agency by the Reagan administration. In the first two years of his administration, Reagan worked to dismantle this creature of Carter's technocratic vision and to tranfer its functions to the corporate dinosaurs populating the energy market or to other more market-oriented departments like the Commerce Department. And, it should be noted that, to the extent that the Reagan administration did possess a national energy policy, its production-guided/supply-side ethic promoted coal policies which would have increased sulfur emissions dramatically by virtue of their call for greater burning of coal for electricity production in the United States.[28]

At the systemic level, one important implication of these developments is that, although the institutionalization of environmental protection into a federal agency had taken place comparatively early in the United States, "the growth of new bureaus [like EPA] does not solve the problems for which they were created."[29] Instead, as a consequence of the highly politicized nature of state institutions of late capitalism, administration typically gives way to politicization and "the state faces an administrative crisis of which its much publicized fiscal crisis is only a part."[30] In other words, the bureaucratization of environmental protection--initiated in the Nixon administration in part to diffuse anti-war fervor--saddled its administrators with all of the political problems and dilemmas from which it was supposed to be shielded by shifting this function from the realm of politics to the realm of administration.

The strategy chosen by the Reagan administration to alleviate or overcome this administrative crisis was two-pronged in nature: the creation of another layer of bureaucratic decisionmaking within the Executive Branch and the imposition of rigidly applied and covertly ideological cost/benefits procedures for assessing federal regulations. In a manner quite contrary to the rhetoric of political decentralization, key architects of the Reagan White House moved quickly in the early months of his first term to establish a decisionmaking framework organized around the model of the administrative presidency.[31] Among other things, this innovation involved the formation of a set of cabinet councils populated by key department and agency administrators and presidential aides responsible for resolving interagency policy and budgetary issues. As the preceding analysis of the rejection of the Ruckelshaus acid rain proposal demonstrates, this new centralizing mechanism at the same time provided another forum for reduplicating policy stalemate at higher levels while it ultimately obstructed the process of policy resolution which had been prepared for at lower levels. That is, the constitution of the Cabinet Council for Natural Resources and the Environment clearly militated against a compromise control policy, which Ruckelshaus and his assistants had constructed out of the political ground shaped by legislative pressure and public support largely mobilized by professional environmental groups and other elements of the pro-control alliance.

The application of cost-benefits analysis to the Ruckelshaus proposal also undercut the case for even this modest, experimental program of acid rain mitigation. The use of this managerial technique in the Reagan administration had its origins in Executive Order 12291 issued in February 1981. The purpose of this order was to require all major proposed federal rules to be subjected to a "regulatory impact analysis" involving a detailed itemization and comparison of expected costs and benefits of each proposed regulation.[32] In the process of fostering a climate for administative deregulation by slowing agency rule-making, it greatly enhanced the role of OMB in regulatory policy by giving supervisory authority over the review process to a new Office of Information and Regulatory Affairs within the budget and management office. Yet, as many critics of cost-benefits analysis have observed, cost-benefits procedures are prone to ideological biases and political judgments especially in politically-sensitive areas because "precise analyses of the benefits of proposed health and environmental regulations remain beyond the capability of state-of-the-art scientific and economic knowledge" and because "even some major costs, such as a regulation's effect on innovation or productivity, are essentially unquantifiable."[33]

Yet another cause fostering the reconsolidation of policy immobilism on acid rain control stemmed from the shift in rationale for postponement by the broader anti-control forces from the scientific argument to the economic argument. Specifically, as the position of scientific agnosticism on the nature and effects of acid rain lost its credibility in the public domain, the regional dispute over the employment and economic impacts of various control strategies assumed new importance and helped to stymy a political settlement. Regional differences had long fueled the debate over acid rain control legislation and complicated the contours of any legislative enactment mandating some form of abatement policy; and labor concerns over possible job losses had been frequently voiced by spokesmen of the United Mine Workers union.[34] Now key politicians seized upon the employment-economic issue in an effort to stave off adoption of acid rain control.[35] Given their narrow outlook on acid rain issues and their consequent failure to forge an environmentalist/labor alliance by combining a strong environmentalist program with employment concerns, the pro-control forces were simply caught off-guard and unprepared for this switch in rationale for policy delay by their opponents.

Despite the nascent consensus within the scientific community on the need for a political decision on some control approach and the pervasive political pressure being brought to bear on legislative access points in the political system, no control proposal was forthcoming from the administration at the end of October 1983. And when the president delivered his State of the Union message to the Congress in January 1984, he announced a willingness on the part of his administration to double research on the problem of acid precipitation but not to take any initiative on a control program for alleviating the problem.[36]

CONCLUSION

Much of the literature on policy immobilism has been linked to conservative or technocratic critiques of the supposedly growing "ungovernability" of constitutional democracies.[37] This chapter, by contrast, has sought to examine the unfolding policy stalemate over acid rain control against a background informed by a critical theoretical perspective. A later chapter will attempt to elucidate more concretely the political ecological framework within which that critical perspective is situated.

For now it will suffice to underscore the irreconcilability of a policy of acid rain control recognizing acid rain as a genuine threat to the ecosystem with the continuing operation of interest-group politics and the prevailing lure of administrative mechanisms of policy resolution in the United States in its late capitalist phase. That is, the interest-group syndicalism of American politics, coupled with the centralizing administrative impulse of executive reorganization in every presidency since Roosevelt, have seriously hobbled efforts to assert ecological priorities in a political economy driven by the pursuit of private profit and economic growth and guided by a view of nature as an object of exploitation. Despite inroads made by the environmental movement in the late sixties and early seventies to educate the public as to the costs to the environment of the so-called "negative externalities" in the industrial production process and to tame the "no limits to growth" mentality of corporate capitalism by lobbying for legislation to internalize these costs, these regulatory checks on the exploitation of nature and on unfettered economic growth have been only weakly institutionalized and their rationale poorly grounded ideologically by American environmentalists.[38] Consequently they have remained vulnerable to countermobilization by powerful economic and political elites in times of economic crisis through the skillful manipulation of antiquated economic symbols and the promotion of shopworn political nostrums.

The major conclusion of this chapter then is that environmentalism will remain a focus of diffuse public support and an interest of professional environmentalist groups capable of imposing costly modifications, but not fundamental reconstruction, of the political economy of corporate capitalism in the United States so long as the interest-group liberal "rules of the game" are taken as given and so long as the favored response to an underlying structural crisis is administrative and not political. By sharing these premises with those in the policy debate with whom they are contending and by advancing technological fixes and regulatory palliatives, environmentalists and other components of the pro-control alliance risk bringing the debate over acid precipitation control to policy closure far short of confronting the level of ecological hazards which threaten the health and equilibrium of the environment. The possiblity of significant acid rain control awaits the transformation of the political tactics and ideological horizons of opposition forces to environmental degradation beyond environmental doomsaying and interest-group reformism.

NOTES

1. Philip Shabecoff, "Acid Rain Options to Be Listed Soon," **New York Times**, September, 1, 1983, p. 14.
2. For an historical review of the acid rain phenomenon, see Ellis Cowling, "Acid Precipitation, in Historical Perspective," **Environmental Science & Technology**, Vol. 16 (February 1982), pp. 110A-123A.
3. This act is reproduced in its entirety in U.S. Senate, Committee on Energy and Natural Resources, **Acid Precipitation and the Use of Fossil Fuels**, 97th Cong., 2nd Sess., August 19, 1982 (Washington, D.C.: USGPO, 1982), pp. 2-4.
4. See John E. Carroll, **Environmental Diplomacy: An Examination and A Prospective of Canadian-U.S. Transboundary Environmental Relations** (Ann Arbor, Mich.: University of Michigan Press, 1983), esp., ch. 11; John E. Carroll, **Acid Rain: An Issue in Canadian-American Relations** (Washington,D.C.: Canadian-American Committee, 1982); Ross Howard and Michael Perley, **Acid Rain: The North American Forecast** (Toronto, Canada: House of Anansi Press, Ltd., 1980); Gregory Wetstone and Armin Rosenkranz, **Acid Rain in Europe and North America:**

National Responses to an International Problem (Washington, D.C.: Environmental Law Institute, 1983), chs. 13 and 14; Jon Luoma, **Troubled Skies, Troubled Waters: The Story of Acid Rain** (New York: The Viking Press, 1984), **passim;** and "The Politics of Acid Rain [Special Issue]," **Alternatives: Perspectives on Society and Environment,** Vol. 11 (Winter 1983).
5. See Boyd Keenan, "Acid Rain Policy: Comparing Canadian and U.S. Approaches," a paper prepared for the 1984 annual meeting of the American Political Science Association convention, Washington, D.C., August 30-September 2, 1984.
6. See Bruce A. Ackerman and William T. Hassler's case study of the Clean Air Act--**Clean Coal/Dirty Air** (New Haven, Conn.: Yale University Press, 1981).
7. See Irvin C. Bupp, Jr., "Energy Policy Planning in the United States: Ideological BTU's," in Leon Lindberg, ed., **The Energy Syndrome** (Lexington, Mass.: D.C. Heath and Company, 1977), pp. 285-324; Michel Crozier, **The Stalled Society** (New York: Viking Press, 1973); Alan Wolfe, **The Limits of Legitimacy** (New York: Basic Books, 1977), pp. 261-271; and Claus Offe, "Ungovernability: On the Renaissance of Conservative Theories of Crisis," in Jurgen Habermas, ed., **Observations on "The Spiritual Crisis of the Age,"** (Cambridge, Mass.: The MIT Press, 1984), pp. 67-88.
8. See Wolfe, **The Limits of Legitimacy,** pp. 261-271.
9. "Environmentalists Won't Vanish Udall Tells Coal Group," **Lexington Herald-Leader,** June 18, 1983, p. D8; and Mary Douglas and Aaron Wildavsky, **Risk and Culture: An Essay on the Selection of Technological and Envioronmental Dangers** (Berkeley, Cal.: University of California Press, 1982), pp. 126-151.
10. Alan Wolfe, **America's Impasse: The Rise and Fall of the Politics of Growth** (New York: Pantheon Books, 1982).
11. Bupp, "Energy Policy Planning in the United States," in Lindberg, ed., **The Energy Syndrome,** pp. 285-324.
12. Otto Kirchheimer, "The Transformation of the West European Party Systems," in Joseph La Palombara and Myron Weiner, eds., **Political Parties and Political Development** (Princeton, N.J.: Princeton University Press, 1966).
13. Geoffrey Norman, "The Acid Rain Wars," **Esquire,** March 1983, pp. 242-243.
14. Rochelle L. Stanfield, "Regional Tensions Complicate Search for an Acid Rain Remedy," **National Journal,** Vol. 16 (May 15, 1984), pp. 860-863.
15. Senate, Committee on Energy and Natural Resources, **Acid Precipitation and the Use of Fossil Fuels,** op. cit., pp. 1222-1281 (Oppenheimer), 1480-1513 (Mohnen); and "A Debate: Are Enough Data in Hand to Act Against Acid Rain?" **New York Times,** November 14, 1982, p. 20E.
16. Richard Klein, quoted in Stanfield, "Regional Tensions Complicate Search for an Acid Rain Remedy," p.

861. See, as well, Gene Likens' remarks in: Woodson Emmons, "Acid Rain is Called a Problem of Politics," **Lexington Herald-Leader,** May 1, 1984, pp. B1 and B3. A key issue in the interplay between policy and science is illuminated by J.L. Regens' essay, "Acid Rain: Does Science Dictate Policy or Policy Dictate Science?" in Thomas Crocker, ed., **Economic Perspectives on Acid Deposition Control** (Boston: Butterworth Publishers, 1984), pp. 5-19.

17. See, e.g., Dudley Clendinen, "Concern on Acid Rain Extending to Public Health," **New York Times,** June 29, 1983, pp. 1 and 8; Bette Hileman, "Acid Rain: A Rapidly Shifting Scene," **Environmental Science & Technology,** Vol. 17 (September 1983), pp. 401A-405A; Eliot Marshall, "Acid Rain, a Year Later," **Science,** Vol. 221 (July 15, 1983), pp. 241-242; and Larry Tye, "Acid Impasse is Finally Dissolving," **Louisville Courier-Journal,** September 19, 1983, pp. 1 and 8.

18. Clendinen, "Concern on Acid Rain," p. 8; "Environmental Update: An Important Value," **Public Opinion,** February/March 1982, pp. 32-37; Philip Shabecoff, "Environment Makes Comeback as Issue," **Lexington Herald-Leader,** May 8, 1983, pp. D2-D4; and "Nation Troubled Over Pollution, Ruckelshaus Says," **Lexington Herald-Leader,** June 23, 1983, p. A6.

19. Martin Crutsinger, "U.S. Report Says Pollution is Likely Cause of Acid Rain," **Louisville Courier-Journal,** June 9, 1983, p. A1; Executive Office of the President, Office of Science and Technology Policy, "Interim Report from OSTP's Acid Rain Peer Review Panel," (Washington, D.C.: Press Advisory, June 28, 1983), mimeo.; and National Academy of Sciences, **Acid Deposition: Atmospheric Processes in Eastern North America** (Washington, D.C.: National Academy Press, 1983).

20. Bette Hileman, "1982 Stockholm Conference on Acidification of the Environment," **Environmental Science & Technology,** Vol. 17 (January 1983), pp. 15A and 18A.

21. Mike Brown, "Coal Industry's Stand Against Restrictions on Acid Rain Crumbles," **Louisville Courier-Journal,** August 29, 1983, pp. A1 and A8; "The Washington Connection: The Acid Rain Lobby Picks Up Steam," **Fortune,** Vol. 107 (May 30, 1983), pp. 33 and 36; and Larry Tye, "Acid Impasse is Finally Dissolving," p. 8.

22. Commoner, **The Closing Circle: Nature, Man & Technology** (New York: Bantam Books, 1974), pp. 204-205.

23. Cited by Norman, "The Acid Rain Wars," p. 243.

24. Philip Shabecoff, "Ruckelshaus Says Administration Misread Mandate on Environment," **New York Times,** July 27, 1983, pp. 1 and 9.

25. See Shabecoff, "Program for Control of Acid Rain Emerging at Environmental Agency," **New York Times,** pp. 1 and 24.

26. Philip Shabecoff, "Ruckelshaus Puts Off Plan to Curb Acid Rain," **New York Times,** October 23, 1983, p. 27.

27. Wolfe, **The Limits of Legitimacy**, pp. 264-265.
28. For a critique of the Reagan energy plan, see Ernest J. Yanarella and Herbert G. Reid, "The Politics of Energy," **Fellowship**, Vol. 47 (July/August 1981), pp. 12-13, and 29.
29. Wolfe, **The Limits of Legitimacy**, p. 264.
30. **Ibid.**
31. For an overview of this presidential model, see generally Richard Nathan's **Administrative Presidency** (New York: John Wiley & Sons, 1983), as well as Robert Bartlett's "The Budgetary Process and Environmental Policy," in Norman J. Vig and Michael E. Kraft, eds., **Environmental Policy in the 1980s: Reagan's New Agenda** (Washington, D.C.: CQ Press, 1984), pp. 121-139.
32. Norman Vig and J. Clarence Davies explore the administrative background and political repercussions of this executive order in: Vig, "The President and the Environment: Revolution or Retreat?" in Vig and Kraft, eds., **Environmental Policy in the 1980s**, pp. 89-90; and Davies, "Environmental Institutions and the Reagan Administration," in Vig and Kraft, pp. 149-150.
33. Davies, p. 150. The role and limits of cost-benefit analysis as a component of policy evaluation of alternative acid rain controls have been explored with insight and thoroughness by Thomas Crocker and James L. Regens: Crocker, "Conventional Benefit-Cost Analyses of Acid Deposition Control are Likely to be Misleading," in Peter Gold, ed., **Acid Rain: A Transjurisdictional Problem in Search of a Solution** (New York: Canadian-American Center, State University of New York at Buffalo, 1981); and Crocker and Regens, "Acid Deposition: A Benefit-Cost Analysis: Its Prospects and Limits," in **Environmental Science & Technology** Vol. 19 (February 1985), pp. 112-116.
34. See Robert Lever, "Acid Rain Bills Threaten Miners' Jobs, Study Says," **Louisville Courier-Journal**, July 10, 1982, pp. A1 and A4; Stanfield, "Regional Tensions Complicate Search for an Acid Rain Remedy," p. 862.
35. See Stanfield, "Regional Tensions Complicate Search," pp. 861-863; Steven Roberts, "Who Will Bear the Costs?" **Environment**, Vol. 26 (July/August 1984), pp. 25-32; and U.S. Congress, Office of Technology Assessment, **Acid Rain and Transported Air Pollutants: Implications for Public Policy** [OTA-0-204] (Washington, D.C.: USGPO, June 1984), pp. 140-146, 190-206.
36. Joseph A. Davis, "Selected Increases Proposed in Environmental Spending," **The Congressional Quarterly**, February 4, 1984, pp. 198-199.
37. See, for instance, the Trilateral Commission, **The Governability of Democracies**, Michel Crozier, Samuel P. Huntington, and Joji Watanuki, rapporteurs (New York: Trilateral Commission, 1975).
38. Consider William Ophuls' treatmemnt of the environmentalist struggle to compel industry to internalize these "external costs" and of the

implications of that partial policy success--Ophuls, **Ecology and the Politics of Scarcity** (San Francisco, Cal.: W.H. Freeman and Company, 1977), pp. 172-174, 176-180.

3
Public Opinion and the Environment: The Problem of Acid Rain

*Phillip W. Roeder
and Timothy P. Johnson*

INTRODUCTION

Acid rain is one of a number of environmental problems that has recently become a hotly debated issue on the governmental agenda of this and many other Western nations. A number of economic, scientific, and political factors help explain why acid rain has emerged as a prominent and controversial public policy issue. Few researchers, however, have focused on public opinion as a major factor in the acid rain debate. Indeed, not only would some observers raise doubts about the public's role in influencing decision-makers on broad issues of public policy; many would question whether the public has even an elementary understanding of the scientific uncertainties, economic trade-offs, and political controversies of this complex environmental issue.
The precise role of public opinion in shaping public policy in our American democracy is ambiguous and controversial. Establishing the links between mass opinion and elite decision-making, and more importantly, the direction of these possible relationships is problematic at best. This difficulty is compounded when dealing with highly technical policy issues such as air or water pollution. Such issues have emotional and rational components which are virtually inseparable. The issues affect jobs and livelihood for large numbers of individuals; they affect the extent to which large regions prosper or decline; and they affect the quality of life of this and other nations. In contrast, these environmental issues are not basic moral issues like abortion or school prayer that eventually turn on one's moral or political values. Instead, they involve a major scientific or rational component which we Americans have come to expect will be resolved by rational, objective procedures rather than be settled by emotional or self-interested reactions.

The purpose of this chapter is to analyze public opinion on the problem of acid rain. In order to put issues of public opinion on acid rain into a broader framework for analysis, some basic findings and useful generalizations about public opinion will be reviewed.

First, there is substantial evidence that few people have opinions on most of the public problems and issues currently being considered and debated by public officials. After citing evidence concerning the high proportions of people who profess "little interest" in presidential elections, and the low proportions of people participating in the electoral process, Hennessy (1981: 14) concludes that "the evidence is undeniable that voters today do not all have an interest in public issues." Erickson and Luttbeg (1973: 26) suggest that based on the public's low level of political information, "many Americans do not follow current events closely enough to develop concrete opinions on topical political issues." They argue that many respondents to surveys provide answers to specific questions more because of their unwillingness to appear ignorant or uninformed than because they have thought about the issue and formed an opinion prior to the interview (Erickson and Luttbeg, 1973: 29).

This widespread existence of "doorstep opinions" is thought to be a major factor in a second generalization about American public opinion. These "pseudo-opinions" are so weakly held or even non-existent that large numbers of people will reverse their opinions from one interview to another at a later time. It should not be surprising then that **public opinion on a number of issues tends to be unstable and shifting.** Since many opinions are so weakly held and of little concern to respondents, the opinions sometimes fluctuate substantially from one opinion poll to another. For example, an ambitious study by Converse which surveyed panels of respondents over time found a high degree of temporal instability in people's opinions on a number of social and political issues (Erickson and Luttbeg, 1973: 29).

Another factor that relates to the "doorstep opinion" explanation for instability is the extent to which opinions can be manipulated by question-wording or sequencing of questions on a survey instrument. The public opinion literature is replete with examples of how "pseudo-polls" use crude, biased question-wording to elicit desired responses. Yet, it has been demonstrated that even subtle, relatively objective differences in question-wording can have noticeable effects on responses to survey questions.

Questions of instability of opinions and "non-opinions" lead to another generalization--namely, that **people frequently express conflicting and contradictory** opinions on a number of issues. A substantial body of research on public opinion explores the degree to which mass belief systems are "constrained," ideological, and/or consistent. A long tradition of research shows that few people demonstrate consistent liberal or conservative positions on most issues. Only recently has research suggested that Americans may be becoming more consistent in their opinions (Monroe, 1975: 158; and Erickson and Luttbeg, 1973: 86). Erickson and Luttbeg (1973: 73) indicate that the American public may become more ideological when a presidential campaign such as the Johnson-Goldwater contest of 1964 fields candidates who identify themselves--and who are identified by others--in strongly ideological terms.

Finally, one of the enduring problems of survey research is the intensity issue. Asking questions of people is not the ideal way to assess or measure the strength of their beliefs or opinions on an issue. As Bogart (1972: 101) suggests, "public opinion is a twilight zone between the learning of information and overt behavior." A reasonably clear expression of intensity of feeling can often be deduced from overt behavior; a much more ambiguous expression is a response to a question provided to an anonymous source on the other end of a phone line or on one's doorstep.

The point of this cursory review of previous public opinion research is not to suggest that such research is useless or even wrong. It is merely intended to urge caution when interpreting the results of surveys of public knowledge and opinions on an issue as complicated as acid rain.

Dimensions of Public Opinion

There are a number of components of public opinion related to public problems and issues. The basic questions pursued in this chapter are the following:

What do people know about the issue? Are people aware of the basic elements of the issue; that is, at the most basic level, do people recognize the term -- "acid rain"?

Has the level of public awareness changed over time? Issues of public importance tend to wax and wane in salience. For example, some policy issues like the Iranian hostage crisis come into public view quickly, become well-known or extremely salient for a period of time, then recede from public concern (usually when the "crisis" ends). In contrast, other policy issues never become salient public concerns. The latter tend to be arcane, scientific questions that for whatever reasons do not capture the public's attention. Current illustrations of these include questions of genetic engineering and its public implications and the issue of building or not building the "star wars" missile defense system. However, a number of complex, highly technical issues do become known and salient to large proportions of the public. Environmental issues tend to be of this type, and it will be our task to assess the extent to which acid rain is salient to the public.

How much does the public know about the issue? Once we discover the level of awareness about an issue, we next ask the question of degree of knowledge about it. One cannot expect the public to have the training and time needed to acquire detailed knowledge of complex issues (even if they were concerned with the problem). Still, we can assess the extent to which people can go beyond simply naming an issue and assess how much they know of the causes and consequences of the issue in question.

How concerned are people about the issue? Hundreds, if not thousands, of issues and problems compete for public attention. Which of these are viewed as most or least serious issues to be concerned with? For example, do people tend to be more concerned with the issue of toxic waste disposal, the growth of the federal deficit or unsafe drinking water? In very crude ways, the public's intensity of feeling about acid rain can be measured.

Does the public have knowledge of--or opinions on--actions to be taken to reduce the effects of acid rain, how much these actions would cost, and how the costs should be distributed among different demographic or regional groupings of individuals? This final area of concern focuses on policy alternatives or "solutions" to the problem.

Methodological Issues

Although sample surveying is probably the most widely known method of social science research among the general public, a number of problems limit the usefulness of information collected using this approach. In most cases only a limited amount of information concerning the topic of interest can be conveyed by the interviewer to a respondent before asking a question. This often results in oversimplification of an issue. In addition, the more complicated the issue in question, the less informed public opinion is likely to be. In studying acid rain, given the fact that many scientific and policy experts disagree about the basic assumptions of the problem, this problem is a particularly serious one.

For a number of reasons we will evaluate and review public opinion on acid rain based on a number of disparate surveys. In Table 3.1 we have provided a summary of the public surveys we use in this chapter.

TABLE 3.1
Surveys reviewed in this chapter

POLLING ORGANIZATION	AREA SURVEYED	SAMPLE SIZE	SURVEY DATES	SURVEY METHOD
1. The Roper Organization	United States	2000	1/9-23/1982	personal
2. The Roper Organization	United States	2000	7/10-17/1982	personal
3. Resources for the Future (for the Council on Environmental Quality)	United States	1576	1/26-2/9 and 3/24-4/5/1980	personal
4. Louis Harris & Associates	United States	approx. 1250 in each year	1980,81,82,83	unknown
5. ABC News-Washington Post	United States	1516	4/8-12/1983	telephone
6. CBS News-New York Times	United States	1489	4/7-11/1983	telephone
7. Gordon Black Associates (for the NY State Dept. of Conservation & the Ctr. for Financial Mgmt SUNY-Albany)	New York State	601	March 1983	telephone
8. Temporary State Commission on Tug Hill & the NY State Cooperative Extension	Tug Hill area of New York State	802	Oct.-Nov. 1982	mail
9. Cambridge Reports, Inc. (for Union Electric Co.)	Missouri (registered voters only)	700	March 1983	telephone
10. Institute for Social Inquiry, University of Connecticut	Connecticut	500	7/25-31/1982	telephone
11. School of Journalism, University of North Carolina-Chapel Hill (for the National Wildlife Federation of North Carolina)	North Carolina	584	10/1-6/1982	telephone
12. St. Paul Pioneer Press and Dispatch/WCCO-TV/WCCO Radio	Minnesota	1119	1/6-15/1984	telephone
13. University of Kentucky Survey Research Center (for the Kentucky Energy Cabinet)	Kentucky	757	6/6-15/1984	telephone
14. University of Kentucky Survey Research Center	Kentucky	969	11/9-12/6/1983	telephone

TABLE 3.1 (cont'd.)

15.	Canadian Gallup Poll Limited (for Canadian Coalition on Acid Rain)	Canada	1058	1/3-5/1985	personal
16.	Canadian Gallup Poll Limited (for Canadian Coalition on Acid Rain)	Canada	1046	March 1984	personal
17.	The Gallup Poll of Canada	Canada	1039	October 1983	personal
18.	The Gallup Poll of Canada	Canada	1060	September 1980	personal
19.	Canadian Facts (for Tricil, Ltd.)	Canada	514	1983	telephone
20.	Centre de Recherche Opinion Publique, Inc. (CROP)	Canada	2003	November 1984	personal
21.	Centre de Recherche Opinion Publique, Inc. (CROP)	Canada	2009	November 1983	personal
22.	Centre de Recherche Opinion Publique, Inc. (CROP)	Canada	1994	November 1981	personal
23.	Centre de Recherche Opinion Publique, Inc. (CROP)	Canada	1939	November 1980	personal
24.	Dept. of Environment, Halifax, Nova Scotia	Canadian Atlantic Provinces	946	Summer 1979	personal
25.	National Wildlife Federation	Acid Deposition Researchers in the United States	200	9/1983-2/1984	mail

In addition, formal citations, where available, are included in the bibliography. As can be seen, there are national surveys, statewide surveys, Canadian surveys, and local surveys, all of which use varied survey methods and question-wording, and sometimes involve very different goals and objectives. These differences in methods and areas sampled all serve to make comparisons difficult. Collectively, these surveys do however form an interesting mosaic of information which contributes to an understanding of the problem if these limitations are kept in mind.

AWARENESS OF ACID RAIN

Basic awareness of acid rain has been examined in almost all the surveys acquired, and the overall results indicate that awareness increased considerably between 1979 (when the first survey was conducted) and 1984. Only 30 percent of those interviewed in Canada's Atlantic Region were aware of acid rain in 1979; yet, only one year later, 66 percent interviewed in a nationwide survey by the Gallup Poll of Canada reported having heard or read about acid rain. In the United States, surveys conducted by Lou Harris found that the proportion of Americans who were familiar with acid rain doubled in four years--from 30 percent in 1980, to 43 percent in 1981, to 55 percent in 1982, to 63 percent in 1983. The Roper Organization also reported that 55 percent of the Americans interviewed in a January 1982 survey had heard of acid rain. Further analysis of the Roper Poll found respondents in the western part of the country to be most likely to have heard of acid rain (67%), followed by the midwest (60%), northeast (57%), and the southeastern (44%) parts of the nation. A number of statewide surveys since 1982 have also found that majorities have heard about acid rain: 64 percent in Connecticut and 60 percent in North Carolina (1982), 93 percent in the Adirondack region of New York and 74 percent in the rest of the state (1983), and 80 percent in Minnesota and 68 percent in Kentucky (1984).

All surveys which include education and income breakdowns show that both factors have clear and positive associations with awareness of acid rain. The surveys also tend to show that males consistently have higher levels of familiarity with acid rain than do females: 64-48 percent in Roper's U.S. study, 72-60 percent in Gallup's 1980 Canadian survey, 69-52 percent in North Carolina, 76-61 percent in Kentucky, and 87-75 percent in Minnesota. The survey data on age differentials are less clear, tending to show middle-aged respondents most likely and older persons least likely to have heard of acid rain.

PUBLIC CONCERN OVER ACID RAIN AND OTHER ISSUES

Several surveys have asked similar questions dealing with concern over acid rain. A Gallup survey in Canada conducted in March of 1984 asked Canadians the degree of their concern about the problem of acid rain in their country. Forty-one percent were very concerned, while 37 percent were somewhat concerned, 12 percent not too concerned, and 9 percent indicated that they were not at all concerned or did not consider acid rain to be a problem. In July 1982, a National survey in the United States by the Roper Organization found 43 percent very concerned about the effect of acid rain on rivers and lakes, 33 percent somewhat concerned, 13 percent not very concerned, and 4 percent not at all concerned. Overall, 78 percent of the Canadians polled, and 76 percent of those in the U.S. were concerned with acid rain. A 1984 statewide survey in Kentucky found that among those having heard of acid rain (68% of the sample), 20 percent were very concerned with it, 52 percent were somewhat concerned, and 17 and 8 percent were not very concerned and not at all concerned, respectively. While the great degree of concern expressed by people in Canada is not surprising, given the extent to which that country has been damaged by acid rain, it is noteworthy that just as many people in the U.S. as a whole--and almost three-quarters of the population in a state not generally associated with damage from acid rain (Kentucky)--are also concerned with it. In all three surveys, the largest difference in level of concern was related to education levels such that the most educated groups were more concerned about acid rain than the least educated by 81 to 70 percent in Kentucky, 89 to 70 percent in Canada, and by 81 to 72 percent in the U.S.

Several studies have assessed the public's feelings about acid rain by raising questions about the perceived seriousness of the problem. Nationally, a January 1982 Roper survey found that 47 percent of the sample felt that acid rain was at least as serious as other environmental problems, or one of the most serious. Of those in the survey having heard of acid rain, however, 86 percent felt it was at least as serious as other problems. Three statewide surveys asking this question found that the proportion of the entire sample that felt acid rain was a serious problem was 33 percent in Connecticut (1982), 41 percent in North Carolina (1982), and 75 percent in Minnesota (1984). For those respondents showing awareness of

acid rain, these figures were considerably higher--
i.e., 51 percent in Connecticut, 69 percent in North
Carolina, and an overwhelming 94 percent in Minnesota
felt that acid rain was a serious problem. Results of
numerous private surveys released in a speech by
pollster Lou Harris (December 1983) indicated that, by
a majority of 90 to 7 percent, Americans view acid rain
as a serious problem. Results from a survey conducted
by the Centre de Recherche Opinion Publique (CROP,
Inc.) in Montreal show that, in 1980, 69 percent of a
nationwide sample agreed that acid rain was one of the
most serious environmental problems in that country.
Subsequent surveys by CROP found this proportion
increasing to 77 percent in 1981 and 1983, and to 86
percent in 1984. A 1983 Canadian Gallup Poll also
found that 79 percent perceived acid rain as an urgent
problem. The Roper survey showed that perceived
seriousness of acid rain increased with both education
and income, and that males and whites were much more
likely to see acid rain as a serious problem than were
females and blacks (53 to 39%, and 50 to 21%,
respectively).

The seriousness with which the public views acid
rain can also be examined within the context of other
environmental issues. This approach has the advantage
of comparing the extent to which people view acid rain
as a serious problem relative to other public concerns.
An ABC News-Washington Post survey conducted in 1983
asked respondents which environmental problems they
viewed as serious. There acid rain was ranked the
fourth most serious of six problems examined, with 23
percent viewing this as a serious problem. The most
serious problem was unsafe disposal of toxic chemical
wastes (35%), followed by polluted lakes and rivers
(30%), the commercial development of open land (24%),
acid rain, problems with nuclear reactors (20%), and
commercial develoment in residential areas (17%). In a
nationwide survey by Lou Harris, four similar
environmental issues were presented to respondents and
then they were asked how serious each was for the
country as a whole. As with the ABC News-Washington
Post survey, the disposal of hazardous wastes was
viewed as the most serious problem (95%).

Other questions comparing concern with acid rain
with other environmental problems permit acid rain to
be viewed in a larger context. A survey in Canada
conducted by Canadian Facts in the summer of 1983 asked
those interviewed how concerned they were with eight
different environmental issues. With 64 percent of all
respondents expressing a high level of concern, acid
rain was ranked fourth. Issues evoking high concern
from more Canadians were: the disposal of waste from
nuclear power plants (76%), the disposal of hazardous

waste from factories (74%), and drinking water quality (68%). At the next level of intensity were: factory smokestack pollution (61%), redevelopment of farmland (55%), pollution from car exhaust (42%), and the locating of dumpsites for household garbage disposal (45%). A recent Kentucky survey (1984) specifically asked respondents familiar with acid rain whether they were more concerned with this or a number of other environmental problems. This approach similarly yielded a hierarchy of concern with these issues. Here, acid rain was found to rank seventh out of nine problems examined behind the disposal of toxic chemical waste (which was of more concern to the population than acid rain by 90 to 6 percent), unsafe drinking water (80 to 16%), untreated sewage (75 to 20%), preservation of farmland (74 to 22%), nuclear power plants (57 to 34%), and protection of endangered species (50 to 45%). While comparisons using different sets of items are difficult, two environmental problems do appear to be viewed consistently as more serious (and/or of more concern) to the general public than acid rain--the issues of toxic chemical waste and drinking water quality/water pollution. In a 1983 CBS News-New York Times Poll, a national sample was asked which environmental problem was most important. Seventeen percent indicated chemical or toxic waste was most important, while air and water pollution (11%) was considered next in importance. Acid rain was tied for eleventh place on the list with four other environmental issues cited by one percent of the sample as the most important environmental problem. In comparison with other questions reviewed here, this question is most interesting because it is an open-ended item and thus required respondents to volunteer an answer. This low level of concern with--or identification of--acid rain lends support to those who feel that forced response questions may elicit "doorstep opinions" that are not strongly held by the public. The responses to this open-ended question also suggest support for those analysts who argue that the term "acid rain" itself evokes strong negative feelings particularly among those who may know little about the problem. Since most people are conditioned to believe that "acid" is an extremely dangerous substance, anything associated with it is also likely to be quickly perceived as unsafe. The results of the CBS/NY Times open-ended questions suggests that, when people are not given the verbal cue "acid rain", the issue may not be as salient to the public as many closed-ended question responses seem to imply.

Surveys comparing concern with acid rain and with social issues also provide additional insight into how the public views this problem. A July 1982 survey by the Roper Organization asked how concerned people were with eleven topics currently being reported by the news media. The effects of acid rain on rivers and lakes was ranked sixth (with 43%) among the eleven items examined. News stories with higher percentages of the sample very concerned were: crime (74%), chemical waste disposal (65%), accidents at nuclear power plants (54%), Soviet chemical warfare in Afghanistan (48%), and abuse of personal information stored on computers (44%). Topics with which fewer people very concerned were: the deterioration of rail and highway bridges (40%), reports of computer embezzlements (39%), drugs and professional athletes (34%), the deterioration of highways (34%), and the defeat of the Equal Rights Amendment (15%). The Kentucky survey also asked respondents familiar with the subject to choose between acid rain and each of six different national problems. Large majorities selected all six problems as of more concern than acid rain: crime (86 to 10%), unemployment (81 to 15%), the future of the Social Security System (79 to 17%), growing federal deficits (75 to 20%), rising utility rates (75 to 22%), and the threat of nuclear war (74 to 22%). Interestingly, the only issue analyzed in both of these surveys, crime, was ranked at the top of each. More important, though, is that when placed within a broad social context, acid rain becomes only one of many important social and environmental issues competing for the public's attention.

CAUSES OF ACID RAIN

Public knowledge of acid rain has been examined in a number of studies. Knowledge of acid rain is treated separately from the question of word recognition reviewed earlier, because the two are very different approaches to the problem. This point is illustrated by a 1980 survey conducted by the Gallup Poll of Canada. Of the 66 percent of the sample having heard or read about acid rain, only 15 percent were able correctly to relate it to sulphur dioxide or nitrogen oxides. Only 65 percent were able to partially define it, while 8 percent gave incorrect answers, and 12 percent did not know. For the entire sample, 10 percent correctly defined acid rain, 43 percent partially defined it, and 13 percent either incorrectly defined it or did not know. As expected, those with a post-secondary education were more successful in defining acid rain correctly (26% correct), and an additional 50 percent were able to give at least a partial definition. In contrast, only 7 percent of those with a secondary education, and 3 percent with a primary education could give a correct answer, while 43 and 35 percent, respectively, were able to partially define it. Less expected were differences by sex: 16 percent of the males and 3 percent of the females correctly defined acid rain, while 44 percent of the men, and 42 percent of the women interviewed gave partially-correct answers.

A statewide survey in New York done in 1983 found that 79 percent of the respondents in the Adirondacks region of the state--an area generally believed to be receiving significant amounts of acid rain--could give meaningful answers when asked its nature. Only 64 percent of those surveyed in the remaining parts of the state were able to give meaningful answers. A study conducted by Resources for the Future in 1980 found knowledge of acid rain (defined as "polluted rain that harms lakes, land and water; that is like vinegar, etc.") to be relatively low at that time (32% correct and partially correct), compared to a host of other environmental issues. Ranked ahead of acid rain in terms of the public's knowledge were: the Three Mile Island problem (77% correct or partially correct), oil production in the U.S. (63% correct), nuclear power plants (52% correct), sources of air pollution (45% correct), cancer causes in rats (42% correct), and synthetic fuels (42% correct or partly correct). Fewer respondents were able to give correct answers to questions concerning nuclear power plant explosions (31% correct) and the Love Canal problem (25% correct or partly correct).

A number of sources have been implicated in the production of the sulphur dioxide and oxides of nitrogen that most scientists believe are the chief causes of acid rain. Among the most commonly-cited factors are coal-fired power plants, automobiles, and industrial facilities (such as steel mills and smelters). Although a scientific consensus on the causes of acid rain appears to be lacking, public perceptions of the problem are crucial for evaluating and developing solutions to the problem.

There are at least five surveys dealing with public perceptions of the causes of acid rain. Nationally, a Roper poll taken in January 1982 found that, among those having heard of acid rain (55% of sample), 58 percent believed that industrial smokestack emissions were its cause. An additional 29 percent thought that emissions from automobile exhausts were the cause. In a 1984 survey in Minnesota asking a similar question of those having heard of acid rain, 43 percent felt that emissions from coal-burning plants were the primary cause of acid rain, while 18 percent identified industrial pollution as the primary cause. A Harris survey found emissions from chemical plants as the number one cause, totalling 64 percent of the nationwide sample. Emissions from plants and factories that burn coal were ranked second (50%), with 43 percent saying that emissions from plants and factories that burn oil and emissions from cars and trucks are also very important causes. Surveys in the Tug Hill region of upstate New York and in Kentucky also asked people what they believed were major causes of acid rain. In Tug Hill, 67 percent said that coal-fired power plants were a major cause, followed by chemical factories (65%), metal smelters (40%), and motor vehicles (39%). The Kentucky survey found factories to be most often cited (64%) as a major cause. In addition, coal-fired power plants (60%), cars and trucks (42%), and oil-fired power plants (60%) were also selected as major causes by a substantial proportion of those having heard of acid rain. In Kentucky, these questions were asked only of those having heard of acid rain.

Several of these surveys have uncovered some unusual findings. Volcanoes were cited as a major cause by 26 percent of those surveyed in Kentucky and Tug Hill, respectively. In Roper's national poll, 15 percent believed that eruptions from the Mt. St. Helens volcano were the cause. In addition, the Kentucky survey found 34 percent of its sample to believe nuclear power plants were a major cause; Tug Hill, 19 percent also thought that nuclear power plants were involved. These results indicate there are still some misconceptions among the population concerning the sources of acid rain.

CONSEQUENCES OF ACID RAIN

As with the causes of acid rain, there appears to be a limited public understanding of the effects of acid precipitation. Problems typically linked to acid rain include: decreasing the pH level of lakes to the point where fish and plant life are either curtailed or destroyed, damage to forested areas, and the deterioration of buildings in some urban areas. Of those having heard of acid rain in the Minnesota survey, 73 percent felt that lakes, forests, and farmlands were being harmed. In Roper's January 1982 national survey, 56 percent of those who are aware of acid rain believe it kills fish and plantlife in lakes and rivers, and 51 percent believe it damages crops and forests. In Tug Hill, New York, 76 percent said lakes and streams were being harmed, 73 percent said fish were being harmed, and 48, 49, and 44 percent, respectively, said that forests, drinking water, and farm crops were also being harmed. In Kentucky, 76 percent of those having heard of acid rain thought it caused damage to buildings and statues, and 66 percent said it was responsible for fish kills in the Adirondacks. A January 1985 sample of Canadians interviewed by the Canadian Gallup Poll Limited for the Canadian Coalition on Acid Rain provides results consistent with the U.S. studies. When asked the area where acid rain was most serious, 67 percent cited lakes, rivers and waterways. An additional 16 percent said forests, and 6 percent felt that fishing was the area most seriously affected.

The results indicate that individuals have reasonably accurate perceptions of the effects of acid rain. However, as with causes of the problem, substantial minorities also have some misperceptions about these effects. Forty-four percent of Roper's sample implicated acid rain in causing respiratory problems and illnesses among people. In Tug Hill, 38 percent also said acid rain was harmful to human health. While this issue is debatable, no strong evidence has yet been reported directly linking acid rain and human health. In Kentucky, a number of environmental disasters were incorrectly associated with acid rain by some segments of the public, including the death of Lake Erie (45%), the failure of orange crops in Florida (29%), the Three Mile Island disaster (23%), and the toxic waste problems at Love Canal and the (Kentucky) Valley of the Drums (21%). As to the potential benefits of acid rain, the public surveyed shows little uncertainty. Asked bluntly in Tug Hill if acid rain was beneficial, not a single respondent answered affirmatively, although 5 percent said it was beneficial sometimes, and 8 percent said only seldom. More substantially, 71 percent stated flatly it was never beneficial, and 16 percent had no opinion.

REGIONAL IMPACTS OF ACID RAIN

Some attention has been focused on geographic areas which are thought to be most affected by acid rain, such as eastern Canada and the northeastern United States. A recent survey in Minnesota showed that 93 percent of those familiar with acid rain (80% of sample) were aware of the fact that many American states had an acid rain problem. When the 1984 Kentucky survey asked respondents which part of the country they thought was affected most, exactly half of those having heard of acid rain (68% of sample) believed that the northeastern part of the country was most affected, while 22 percent identified the western region of the country, 14 percent Kentucky and neighboring states, and 4 percent southern states such as Florida as the area hardest hit by the effects of acid rain. The same survey found some possible contradictions in the public's perceptions of regional differences. Over 54 percent of those familiar with acid rain agreed that "virtually all of the acid rain the northeastern part of the U.S. is caused by pollution from coal-burning power plants in the midwestern states such as Ohio and Indiana." After implicating this region of the country in contributing to the acid rain problem in the Northeast, another question found that 79 percent of the same respondents said that they thought that most of the acid rain falling on New York came from within its own borders. A statewide survey in New York showed that 73 percent of the residents in the Adirondack Mountains region thought acid rain was a problem in the state compared to 58 percent of those living in other parts of New York. Emphasis upon the damage caused to Adirondack lakes may account for this difference. In Tug Hill, a very similar proportion, 75 percent, thought acid rain was probably a problem there.

Two surveys done in 1984 appear to reflect regional differences in the perceived personal effects of acid rain. In Kentucky, 34 percent of those having heard of acid rain felt they were not at all affected by it, while an additional 33 percent thought they were affected only a little. Sixteen and 5 percent, respectively, said they were affected a fair amount or a great deal. In New York, majorities of those living in the Adirondacks (69%) and the rest of the state (58%) said severe damage from acid rain in the Adirondacks would affect them personally.

Without question, the most important regional aspect of this issue is the transport of acid precipitation across international boundaries. Recognized as a problem in Europe for several decades, this issue has only received widespread attention in North America since the late 1970s. It is generally believed that a sizeable proportion of the acid rain falling on eastern Canada is transported there by prevailing winds from the midwestern section of the United States. Interestingly, at least four different nationwide sample surveys in Canada have examined some aspect of this transboundary problem, while none, to our knowledge, has been undertaken in the U.S. In April 1984, CROP asked a national sample of Canadians who was **more** to blame for the acid rain problem in North America. The U.S. was blamed by 56 percent of the sample, while only 3 percent held Canada responsible. Thirty percent felt that both Canada and the United States were equally to blame. A 1984 Gallup survey revealed that two-thirds of the Canadians interviewed believed that 50 percent or more of the acid rain falling on that country was coming from the United States. The 1984 CROP survey also asked what was the most serious problem in Canadian-U.S. relations. A plurality of 26 percent cited acid rain, followed closely by nuclear arms control (24%). An earlier Gallup Poll (1980) found Canadians to be somewhat skeptical of the success of a recently signed agreement between the U.S. and Canada to curb acid rain and other international air pollution problems. Of those aware of acid rain, only 6 percent expressed confidence that the agreement would be very successful, in contrast to 41 percent who believed it would not be too successful, and 43 percent who felt it would be somewhat successful. Most recently, Gallup (1985) asked Canadians if they believe their country should await U.S. action before beginning its own acid rain clean-up program. Eighty percent answered that Canada should not wait for the United States, while only 11 percent felt that they should. Moreover, although Canadian citizens are unhappy with the U.S. over the issue of acid rain, they are also somewhat critical of the performance of their own country. Gallup's 1985 survey reported that three-quarters of those interviewed felt that Canadian governments had made either no progress (23%) or only a little progress (51%) in solving the acid rain problem. Comparative U.S. survey data remain to be collected.

POLICY IMPLICATIONS: COST OF CONTROLS

Several important policy issues surround the acid rain controversy. Chief among these are: what should be done to reduce the effects of acid rain? Who should be responsible for solving the problem? Who should bear the costs of solving the problem? And should any actions be attempted before more scientific research is conducted? Regardless of what actions are ultimately taken, the public seems to be aware of the costs involved. For example, among the three-quarters of those Kentuckians interviewed who were familiar with the problem, the belief was widespread that it would cost a great deal of money to stop most of the acid rain currently falling. While the above questions are complicated and often interrelated, enough survey research now exists to begin sorting out public preferences about some of the policy choices that must be made.

The most common proposal for reducing acid rainfall involves the installation of devices called scrubbers on the smokestacks of coal-fired power plants and various types of factories. According to many experts, this process can reduce the amount of sulphur dioxide released into the atmosphere. Most surveys dealing with this proposed solution have alerted respondents to the costs involved. In Missouri, when a sample of registered voters were asked about a specific plan to reduce the emissions from coal-burning electric plants, sixty-two percent supported the proposal. When told that this action would cost consumers around $7 billion annually, public opinion became less decisive. When a dollar figure was given, support fell to 47 percent and opposition grew to 50 percent. Asked if they would favor or oppose the proposals if it resulted in an average increase of $300 per year in electricity and other costs, only 9 percent of the sample supported it and ten times as many (90%) voiced opposition. When asked how much they would be willing to pay themselves on a monthly basis to reduce emissions of sulphur dioxide, 30 percent said nothing, 32 percent said five dollars or less, 12 percent said 6 to 10 dollars, and 9 percent were willing to pay more than 10 dollars per month.

When respondents in Kentucky were told that reducing sulphur emissions would increase the costs of electricity, 55 percent of the sample indicated a willingness to pay an additional $25 to $50 a year in utility costs to help reduce acid rain. Only 24 percent were willing to pay an extra $100-$150 per year, and only 8 percent indicated they would accept an additional cost of $200-$300 a year. When informed that increased electricity costs would occur in only certain parts of the country, 79 percent of the sample indicated that they would favor some method of spreading the costs of acid rain controls more fairly. In New York State, 97 percent of those in the Adirondack region and 95 percent of those living in the rest of the state favored pollution controls when cost was not a factor. When informed that the consumer could ultimately have to pay the costs of reducing acid rain, 50 percent in the Adirondacks, and 47 percent elsewhere in the state said an extra $25 per month in electricity bills would be worth the cost. A 1983 Canadian Gallup Poll reported that 86 percent of the public favored placing restrictions on Canadian industries that were responsible for producing acid rain. In the following year, another nationwide sample by Gallup asked how much Canadians would be willing to pay in higher taxes and product costs to clean up acid rain. Thirty-five percent said nothing, 31 percent said $5 per month and 23 percent indicated they were willing to pay $10 or more per month. When this question was again asked in January 1985, the proportion indicating a willingness to pay additional costs to finance a clean-up increased slightly from 54 to 56 percent of the sample. In North Carolina, 89 percent of those believing acid rain to be a serious problem (60% of sample) also felt that acid rain should be more tightly controlled. Asked if they would continue to favor acid rain controls if it meant a "modest" increase in utility bills, 83 percent said yes. As these results demonstrate, support for controls is much higher when specific costs to individuals are not mentioned. As with many public issues, a broad concern for controlling a perceived problem is usually not supported by a willingness to pay more than minimal amounts of out-of-pocket expenses. Generally, support for implementing controls even if it increased utility bills was found to increase with both education (from 71% with 8 years or less of school to 92% of those with at least some college education) and income (from 70% of those with incomes of $10,000 or less to 89% of the respondents earning over $20,000) in this survey.

An alternative to scrubbers for reducing sulphur emissions is the use of low-sulphur coal. For the most part, this option would require eastern and midwestern coal users to import more coal from western states, such as Colorado and Wyoming, which have an abundance of low-sulphur coal. The effect upon employment in eastern states, which mine mostly high-sulphur coal, could be substantial. The only survey examined which deals with this topic was conducted in Kentucky--a state where certain regions can expect to suffer if the use of low-sulphur coal becomes mandatory. When asked if changing to low-sulphur coal was preferable if it were cheaper than the installation of scrubbers, 67 percent of the sample said yes. Yet, of those favoring the switch to low-sulphur coal, only 33 percent would continue to do so if it resulted in lost coalmining jobs in Kentucky. Only 22 percent would still favor switching types of coal if 1,000 jobs were lost, and only 11 percent would favor it if 10,000 jobs were lost in Kentucky.

These results indicate that the public, in principle, favors some of the policy alternatives proposed to lower the smokestack emissions blamed for having a major role in producing acid rain. However, when asked about their willingness to pay more money in taxes, utility bills, the cost of various products, or to sacrifice local jobs, they have been more reluctant to support controls.

To assess public belief concerning the distribution of the burden of the cost of cleaning up acid rain, Louis Harris asked a national sample of Americans about the fairness of four different ways of paying the costs of solving the acid rain problem. The largest majority--73 percent--felt that businesses or individuals who use fuels that contribute to acid rain should bear the costs. Also acceptable to a large majority was requiring shareowners of electric utilities to pay for the clean-up. Smaller proportions of the sample believed it would be fair to require either that electricity users nationally (60%) or that only electricity users in the eastern U.S. (54%) be responsible for the costs of solving the problem. Regional differences on this last proposal were apparent. For instance, only 50 percent of those living in the East and 52 percent of those in the South thought the idea of requiring eastern electricity consumers to pay all of the costs of the clean-up was fair. On the other hand, larger proportions of those in the Midwest (56%) and the West (60%) believed electric consumers in the East, where most of the coal associated with acid rain is burned, should assume the financial burden of solving the problem was fair.

Other surveys have asked respondents specifically who should bear the costs of solving acid rain. In Kentucky, a plurality (42%) of respondents believed everyone should pay for it. Smaller numbers designated industries along (28%), the government (15%), and consumers (2%) as the group that should pay. Given a choice among a number of ways to pay the clean-up costs, respondents in Minnesota familiar with acid rain overwhelmingly supported the idea of charging major polluters a special fee (70%). General taxation was supported by 8 percent of those interviewed, and 9 percent favored some combination of approaches for covering the expense of stopping acid rainfall. Minnesotans were also asked which ways of paying for acid rain clean-ups they would oppose. Over half of the sample, 52 percent, included general tax incentives as something they would oppose.

The role of the government in bearing the cost of reducing acid rain has been examined in Canada. Gallup's 1984 survey asked people if they felt the government should contribute to the cost of reducing the acid rain-causing emissions of industries in Canada if those industries could not afford to themselves. Three-quarters of the sample (74%) said they felt the government should contribute financially and 18 percent opposed this idea. Almost half of the sample (47%) thought that the government should pay at least 50 percent of the cost of the clean-up. As part of the same survey, respondents living in Ontario were asked specifically about the International Nickel Company (INCO) smelter in Sudbury. That plant has been referred to as the single largest source of acid pollution in the world (Sweet, 1982). Sixty-five percent of the Ontario residents indicated that they would approve of INCO, the federal government, and the

Ontario government each sharing equally in the cost of cutting-down the emissions from the smelter. Seventeen percent felt INCO should pay a larger proportion or all of costs, and 4 percent thought a better solution could be found.

The role of government in dealing with the problem has also been examined in New York. Statewide, 72 percent of the population believes that national and state governments should work together to solve the problem. Twelve percent believes the federal government alone should solve the problem, and 7 percent feels state governments should be responsible for dealing with it. In the 1982 Tug Hill survey, a majority of the New Yorkers in that part of the state (64%) thought that the government in general was not paying enough attention to acid rain.

A final policy question in the acid rain debate involves the choice of applying controls on sulfur emissions or postponing such controls until more scientific research has been conducted. The Reagan Administration and many utility companies have supported the position that more research should be done before any controls are introduced. Environmental groups have taken the position that controls on sulphur emissions must be applied quickly if there is to be any hope of saving many of the lakes and forests currently threatened by this problem. In the Tug Hill (New York) survey, 60 percent of the sample indicated that they thought additional research was necessary before any controls were applied. A 1983 University of Kentucky Survey Research Center survey asked a statewide sample of respondents if they believed more research was needed before efforts to solve the problem are made, or whether they felt that enough was already known to begin programs to reduce acid rain. Forty-five percent of the sample thought that programs should be initiated now, while 37 percent felt that more research was needed. The responses varied considerably by education, with the proportion of those wanting programs to be started now ranging from 33 percent of those with 8 years or less of school to 57 percent of those with graduate school experience. Another survey in Kentucky taken the following year asked the same question in a different form, one that emphasized the disagreements among scientists, and found a majority (63%) responding that more research should be conducted first, with 26 percent insisting on controls being applied. Since these surveys were conducted at different times, a number of factors may have contributed to the large increase in the proportion of Kentuckians who felt that more research should be conducted before applying controls on acid rain (37% in 1983, 63% in 1984); however, the change in question-wording illustrates a point made previously that opinions can be changed even by subtle, relatively objective changes in wording. In the latter survey, education was again found to vary with response to this question. Regardless of whether or not Kentuckians supported immediate acid rain controls, 96 percent of those interviewed in 1984 also agreed that scientific research and development efforts on this problem should be continued.

SUMMARY AND CONCLUSIONS

Passing judgment on the public's knowledge and consistency of opinion on acid rain is difficult. Frequently, no anchor or meaningful standard to make judgments exists. Thus, in reviewing these findings, it is tempting to conclude that the public knows little about the causes and consequences of acid rain and that many of these surveys are finding what we referred to previously as "doorstep opinions." This may be the case, but it must be remembered that, in spite of the scientific component of this problem, even the experts are ambiguous about these issues. For example, Record et. al. (1982: 15) states that "in general, there has been no clear quantification of the magnitude of the potential adverse or beneficial impacts of acidic precipitation." On the other hand, these same researchers state: "understanding of the changes brought about by acid deposition is far from complete, but as increasing amounts of data are evaluated, there is general agreement that the effects are, on balance, detrimental to the environment" (Record, et. al., 1982: 26).

Even the best scientific studies leave room for conjecture. Perhaps the most prestigious and publicized review of the scientific evidence was unclear or inconclusive in its findings and recommendations. The National Academy of Science (1983: 5) reports that "evidence exists for long-range transport of pollutants leading to acid deposition, but the relative contributions of specific source regions to specific receptor sites currently remain unknown." Or further (p. 11): "because we cannot rely on current models or analyses of air-mass trajectories, we cannot objectively predict the consequences for deposition in ecologically sensitive areas of changing the spatial pattern of emissions in eastern North America, such as by reducing emissions in one area by a larger percentage than in other areas."

Finally, concerning the national policy implications of its study, the NAS report (p. 24-25) concludes that, if national policy on acid deposition is to be made on the basis of the scientific information currently available, that policy could take several forms, including maintaining the status quo. Other policies might incorporate uniform reductions (rollback) in emissions, might be designed to achieve the maximum possible environmental benefit, or might be carefully engineered to bring risks, costs, and benefits into optimal balance. The different options require scientific and technical information in different degrees of detail. In addition, they all, to one degree or another, must account for uncertainties in understanding. Decisions on almost all issues of public policy--including military affairs, the economy, and social welfare no less than environmental issues--are routinely made in light of uncertainties in knowledge. Provided uncertainties are taken into account, sufficient information is available for deciding what, if anything, to do about acid deposition.

A concerned and interested citizen then is unlikely to gain much guidance or to reduce his or her uncertainties on acid rain by analyzing the judgments of scientific experts. At the same time, former EPA Administrator Ruckelshaus is reported to have said in a speech that "if acid rain controls were cheap, there'd be no dispute about the science" (Dysart, 1984: 6).

Given these uncertainties, what can be stated with relative confidence about public opinion on acid rain? Awareness of the problem of acid has increased dramatically over a brief period of time. Moderately large proportions of people are aware of the problem and, as would be expected, these proportions vary significantly by region and by certain demographic characteristics. The extent to which the public views acid rain as a serious problem varies considerably based on the study and the methods. Approximately 70% of the public views the problem as serious, but when compared to other issues, acid rain tends to fall somewhere below the top issues of public concern. Much depends on question-wording and to what issues acid rain is compared.

Harris national survey results of Americans tend to place acid rain higher as a subject of concern than other surveys, even those of Canadians. As would be suggested by previous research on public opinion, people are not generally well-informed of the causes and consequences of acid rain. Even when less-than-stringent standards are used to judge knowledge, only moderate proportions of the public can give meaningful or correct answers about acid rain.

In spite of some gaps in scientific knowledge, the public does appear to be reasonably informed about the various actions that have been proposed to deal with the problem, and the costs of these proposals. In spite of willingness to support regulation or controls in principle, the public tends to back away from such actions when a personal price-tag is attached to the action. Americans and Canadians appear willing to pay small amounts themselves to control emission, but when amounts of $100 a year and above are mentioned, these individuals lose their enthusiasm for controls. This certainly demonstrates self-interest if not rationality.

REFERENCES

Bogart, Leo. **Silent Politics: Polls and Awareness.** New York: John Wiley and Sons, Inc., 1972.

Dysart, Benjamin C. President of the National Wildlife Federation. Statement at Hearing on Acid Rain Control Legislation Before the Committee on Environment and Public Works, U.S. Senate, February 9, 1984.

Erickson, Robert S., and N. Luttbeg. **American Public Opinion.** New York: John Wiley and Sons, Inc., 1973.

Harris, Louis. Remarks to the Coalition of Northeastern Governors, Meadowlands, N.J., December 4, 1983. Louis Harris and Associates, Inc., 1983.

Hedges, Roman, and P. Wissel. "Public Perceptions of Acid Deposition: A Survey of New York State Residents." [Exhibit 5] in Donald J. Reeb, ed., **The Economic Impact Study of Acid Precipitation.** Center for Financial Management. Rockefeller College, State University of New York, Albany, New York, 1983.

Hennessey, Bernard. **Public Opinion.** Monterey, Cal.: Brooks/Cole Publishing, 1981.

Kentucky Energy Cabinet. "Public Opinion Poll: What Kentuckians Think About Acid Rain." Kentucky Energy Cabinet, P.O. Box 11888, Lexington, Kentucky, August, 1984.

Mitchell, Robert Cameron. "Public Opinion on Environmental Issues." Washington, D.C.: Council on Environmental Quality and Resources for the Future, 1980.

Monroe, Alan D. **Public Opinion in America.** New York: Harper and Row, Publishers, Inc., 1975.

National Academy of Sciences. **Acid Deposition: Atmospheric Processes in Eastern North America.** Washington, D.C.: National Academy Press, 1983.

Record, Frank A.; Bubenick, David A.; and Robert J. Kindya. **Acid Rain Information Book.** Park Ridge, N.J.: Noyes Data Corporation, 1982.

Sweet, William. "Acid Rain." In H. Gimlin, ed., **Environmental Issues: Prospects and Problems.** Washington, D.C.: Congressional Quarterly, Inc., 1982.

Temporary State Commission on Tug Hill. "Tug Hill Area Acid Rain Public Opinion Survey." State Office Building, Watertown, New York, May, 1983.

Part II

Policy Issues and Alternatives

The allocation of the costs and benefits of various acid rain control strategies must be satisfactorily addressed if the intractability of the acid rain control policy debate is to be overcome. The debate over distributional issues, however, has been intensified by an asymmetrical relationship between environmental benefits and economic costs. As Larry Regens and Robert Rycroft highlight in their introductory chapter to this part, the environmental benefits of an acid rain control program (e.g., increased surface water akalinity and improved visibility) are often perceived as abstract or merely subjective and thus frequently difficult to convey to policymakers and to the general public. Consequently, the social valuation of environmental benefits appear to be difficult--though not, in principle, impossible--to define. On the other hand, the economic costs of such a program (e.g., the loss of 89,000 jobs and a 20 percent increase in electric utility rates) can be easily defined quantitatively and can be easily communicated.
Regens and Rycroft also demonstrate how the complexity and incomplete status of the science of acid rain, including its environmental effects, compound the difficulty of decisionmaking because risks and culpability--important issues in designing a control strategy--cannot be definitively assigned. Nevertheless, proponents of acid rain legislation argue that enough is known to design and implement an effective control program. Yet, as efforts to design such a strategy proceed, policymakers are confronting the problem of designing a strategy which is both effective in terms of achieving specific environmental benefits and appropriate in terms of allocating costs according to socially and politically acceptable criteria. Working within the general framework of issues outlined in Regens and Rycroft's chapter, the other chapters in this part strive to refine our understanding of these issues and to offer alternative policy avenues for breaking through the prevailing policy impasse.

Two key strategic issues which must be confronted in the fashioning of a program for acid rain control are: whether or not cost allocation should emphasize equity or efficiency; and whether distributive issues are best resolved in terms of market economics or within a more overtly political context. In other words, the issues may be reduced to: "Who pays?" And "how do we decide?" Regarding the former, financing the costs of reducing SO_2 emissions is of growing importance in the quest for a resolution of the current policy stalemate. Yet, a policy which provides financial mechanisms (e.g., a trust fund) for the costs of clean-up is at odds with the principle that the "polluter pays," which has been the foundation of current envionmental policy. Given the admittedly high costs required for virtually any strategy of acid rain control, practical and political considerations may compel policymakers to confront the limits of applying traditional policy responses to this environmental problem.

Proponents of acid rain policy based on the "polluter pays" principle argue that such a program is not really imposing "new" costs; rather, they say that such a program merely involves shifting the burden of costs presently not being borne by the region suffering the effects of acid deposition to the region responsible for the damage. In other words, one region of the country (i.e., the Midwest) is enjoying the benefits of not having to control powerplant emissions, while another region (i.e., the Northeast) bears the environmental and public health costs. One difficulty is that assignment of costs to the "polluter" presupposes that a particular source can be identified and that culpability can be unambiguously established. In the case of acid deposition where the mixing and chemical transformation of pollutants in the atmosphere occurs over an entire region (i.e., eastern North America), the applicability of the "polluter pays" principle may be quite limited. The reason for this is that this rule of thumb does not necessarily imply that costs can allocated only where discrete sources can be isolated. It could also apply to a source region, such as the Midwest, and/or to generalized sources, such as coal-fired powerplants with certain characteristics. Yet, if such a control strategy is to be politically defensible the projected economic costs at a minimum must bear some reasonable relationship to the expected environmental benefits.

The chapters by Glenn Gibian and David Streets address this issue and, in the process, offer critiques of recent legislative proposals for acid rain control on the basis of their distributional consequences. Gibian, for example, seeks to determine the extent to which most of the proposed regulatory control strategies can be expected to yield environmental returns commensurate with their economic price-tags. On the basis of computer modeling results, he suggests that strategies which impose large emissions reductions on midwestern and Appalachian sources will yield only small improvements in precipitation acidity in the northeastern United States. One important reason for this result, as Streets points out in his chapter, is that it is not the case that the emissions of all states contribute the same amount of deposition at sensitive receptor areas. A policy which ignores this distinction and assumes additionally that all locations are equally sensitive to acid deposition will not be cost-effective.

On the issue of who decides, several policy recommendations are forthcoming in various chapters in the second part of this volume. As many policy analysts have observed, despite the formal advantages of a market-based regulatory program (e.g., administrative flexibility and efficiency) and the attractiveness of the economic logic underlying such an approach, a variety of powerful political factors severely limit its practicially. As an alternative, Webber proposes a "quasi-market regulatory scheme" which includes economic incentives, an emissions tax, and a central role for states in achieving mandated sulfur dioxide emission reduction targets. Such a "state-based" program, Webber contends, would yield results which are both equitable and efficient.

Streets takes a different tack on the analysis of efficiency issues. Based on work conducted at the Argonne National Laboratory, he examines acid rain control schemes in terms of several efficiency parameters in order to illustrate how "optimal configurations of emissions reductions can be selected to improve the efficiencies of a control strategy." He advances the argument that achieving "real world efficiencies" will entail the application of economic incentives and penalties to alter the economic choices confronting electric utilities and other major emissions sources so that "if the sources make rational economic choices then the emission reduction goal should be achieved efficiently." His analysis also suggests that the implications of control strategies for "regional coal use" may represent an outer limit to control strategy optimization because "the greater the efficiency of a [strategy] the greater the fuel switching that occurs, and the greater the diruption of exiisting patterns of coal supply." In principle, Streets' analysis suggests, a properly designed, targeted reduction strategy could minimize disruption in traditional coal markets while maximizing efficiency.

Some opponents of acid rain legislation have argued that authorities provided under the Clean Air Act are sufficient to meet the acid rain problem. Additional legislation, they claim, would be too disruptive, expensive, and redundant. Whether true or false, these arguments underscore the importance of considering the relationship of acid rain control strategies to aspects of the Clean Air Act (particularly, the National Ambient Air Quality Standards and the New Source Performance Standards) in any credible effort to construct an effective acid rain control program.

In their chapter, Larry Parker and John Blodgett analyze the "present status of the [Clean Air Act] vis-a-vis the acid precipitation problem." There, they point out that there are significant incongruities between the acid deposition phenomenon and the scope, focus, and structure of the Clean Air Act, especially its emphasis on ambient concentrations rather than on total pollutant loadings. Indeed, as they argue, the act may actually have played a role in exacerbating the problem by allowing the use of tall smokestacks as a means of complying with local air quality standards.

Their analysis of the role of the New Source Performance Standards (NSPS) as a long-term emission reduction program is particularly important to the consideration of policy alternatives in the light of the recent attention given to "the issues of how to abate emissions from existing facilities and whether the natural turnover of facilities will proceed fast enough to make a massive retrofit control program unnecessary." Resolution of this issue depends upon a number of economic and other factors influencing expectations of future capacity requirements and bearing on utility decisions regarding powerplant turnover rates. Parker and Blodgett's examination of these factors leads them to conclude that SO_2 emissions will probably "**not** continue to decline" without affirmative Congressional action. Such positive legislative action might include significant revision of NSPS, coupled with a retrofit program for pre-NSPS electric-generating facilities, or "government action to shorten [their] lifespan."

One key implication of Parker and Blodgett's analysis then is that the issue that looms large in the background of the acid rain debate is whether or not it is necessary to re-evaluate critically the structure and orientation of the Clean Air Act. Indeed, the proposals offered by these legislative researchers implicitly assume that substantial changes in the nation's approach toward air quality issues may be required to deal with the problem of acid rain. In this context, it is important to note that the acid rain phenomenon can be viewed as a particularly troublesome form of "transboundary air pollution," which, as Parker and Blodgett indicate, is addressed only in a cursory manner by the Clean Air Act. In effect, their analysis provides evidence that ad hoc approaches to some of the fundamental problems of the Clean Air Act do little to rectify these problems. If so, any serious effort to transcend the acid rain policy impasse would seem to lead ineluctably to a fundamental re-evaluation and restructuring of the Clean Air Act itself.

4
Perspectives on Acid Deposition Control: Science, Economics, and Policymaking

James L. Regens
and Robert W. Rycroft

INTRODUCTION

Over the past decade, the phenomenon of acid deposition, commonly referred to as "acid rain,"[1] has been the subject of growing scientific research and media coverage. Concern has been expressed about the possible short-term and long-term consequences of the deposition of sulfuric and nitric acids on aquatic ecosystems, forests, crops, and man-made structures or cultural artifacts--especially in eastern North America and Europe. Such attention has fostered increased public awareness of acid rain's existence. For example, while few Americans identified acid rain as an air quality problem in 1980, a September 1983 Harris poll found that 63 percent of those questioned were aware of acid rain and approximately two-thirds favored stricter controls on sulfur dioxide (SO_2) emissions, one of acid rain's precursors. Almost simultaneously, acrimonious debate about the extent to which acid deposition constitutes an environmental risk requiring prompt regulatory action to mitigate or prevent its effects has emerged within the political process. Disagreement about the merits of various recommendations to reduce substantially the emissions of acid rain precursors into the atmosphere, primarily SO_2 emissions in the eastern U.S., has become a focal point for the reauthorization of the Clean Air Act. As a result, acid deposition has been transformed from a relatively unnoticed area of scientific inquiry into a major environmental issue of the 1980s.

The focus of debate has shifted increasingly away from questions of science to discussions of how to allocate the costs of emissions reduction programs. Because sulfate (SO_2) is the major constituent of acid deposition in eastern North America as well as Europe, advocates of controls have emphasized reducing SO_2 emissions. Yet, the complexity--especially since distant and local sources contribute to acid deposition--in conjunction with the size of control costs make agreement on equitable reduction strategies difficult to achieve. As a result, the controversy surrounding the adoption of potential acid rain control strategies illustrates the problems of environmental policymaking. To understand those problems, requires insights into the science and economics of acid rain as well as the politics, especially interest group and bureaucratic politics, of the issue.

EMERGENCE OF THE ISSUE

Acid rain first emerged as a public policy concern at the 1972 United Nations Conference on the Human Environment in Stockholm. In a case study prepared for the Stockholm Conference, Swedish scientists asserted that precipitation acidity attributable to SO_2 emissions from man-made sources was causing adverse ecological and human health effects (Royal Ministry for Foreign Affairs, Royal Ministry of Agriculture, 1972). Largely in response to the Swedish study, a number of major research efforts were initiated to address the question of acid deposition's causes and ecological effects. By the mid-1970s, declining pH and possible impacts on aquatic and terrestrial ecosystems had been reported in Sweden, Norway, Canada, and the United States (Cowling, 1981 and 1982). Since the mid-1970s other western European countries such as the United Kingdom, the Federal Republic of Germany (FRG), the Netherlands, and Austria have become increasingly more concerned as potential impacts have been identified in both their own and neighboring states (see Vermeulen, 1978; Wright et. al., 1890; and Ulrich, 1982).

The issue's growing prominence has produced attempts to address it in a variety of forums in the years since the 1972 Stockholm Conference. The Organization for Economic Cooperation and Development (OECD) has adopted recommendations for national SO_2 emissions control programs and endorsed attempts to reduce transboundary air pollution (OECD, 1978a; MacNeill, 1983). Parallel efforts were initiated in the United Nations Economic Commission for Europe (ECE) culminating in the 1979 Convention on Long-Range Transboundary Air Pollution.

In North America, the U.S. and Canada have made limited progress in developing a bilateral agreement on transboundary air pollution. In 1978, the two governments established a Bilateral Research Consultation Group on the Long-Range Transport of Air Pollutants to coordinate the exchange of scientific information on acid deposition (Altshuller and McBean, 1979). During the fall of 1978, the Congress passed a resolution calling for discussions and on August 5, 1980, the two governments signed a Memorandum of Intent concerning Transboundary Air Pollution (MOI) as a framework for bilateral negotiations. Formal negotiations, started in the fall of 1981, currently are marked by disagreement over a 1982 Canadian proposal for joint 50% reductions in SO_2 emissions. U.S. officials labeled the idea "premature" and instead urged continued cooperation to enhance scientific understanding of the phenomenon (see Editorial Research Reports, 1982: 63-80; Marshall, 1983).

As a consequence, countries advocating emissions reductions beyond current air quality management requirements have adopted a dual strategy of relying on domestic controls combined with efforts to encourage multilateral agreement on specific reductions under the aegis of the ECE Convention. The environment ministers of Canada, the FRG and eight other western European countries[2] signed a declaration in Ottawa on March 21, 1984, to reduce sulfur emissions at least 30% in the coming decade as a means to pressure neighboring countries--the United States the United Kingdom, and Belgium--to make similar pledges. Although the United States and the United Kingdom declined to endorse such a target, the September 1984 meeting of the Convention's Executive Body (EB) did, in fact, produce an agreement to hold follow-up discussions on incorporating the 30% cut into the legal framework of the Convention and an additional eight countries agreed to join the "30% Club."[3] At least, symbolically, this move may reinforce the arguments of proponents of taking further control measures now.

At the same time, within the U.S. political system, legislation proposing substantial reductions in SO_2 emissions has been introduced in each Congress since 1981. In August 1982, the Senate Environment and Public Works Committee reported out a bill containing comprehensive Clean Air Act amendments, including an eight (8) million ton SO_2 rollback with offsets for new sources other than coal-conversions in the 31 eastern states (S. 3041). On March 7, 1984, the Environment and Public Works Committee voted 14-2 in favor of an even tougher acid rain bill requiring a 40% reduction in SO_2 emissions by 1994 (S. 768).[4] A stalemate has existed for the past few years in the House Energy and Commerce Committee over proposed acid rain controls, although the Subcommittee on Health and the Environment considered legislation. For example, on June 23, 1983, H.R. 3400 was introduced to achieve a 10 million ton reduction of SO_2 emissions in the 48 coterminous states by January 1, 1993 (the Waxman-Sikorski proposal). The Waxman-Sikorski bill would concentrate the reductions on 50 utility power plants having the highest 1980 SO_2

emissions and require those plants to install flue gas desulfurization technology (FGD) by 1990.[5] It also would establish a fee on all non-nuclear electricity generation starting in 1985 to reimburse utilities for 90% of the capital costs of installing pollution abatement technology on the 50 specified plants. By a 10-9 vote in mid-1984, the subcommittee failed to approve H.R. 3400, effectively ending any possibility of further control measures being adopted during the 98th Congress. Thus, passage of acid rain legislation mandating additional emissions reductions will not occur until sometime in 1985 at the earliest. And, because support for the various proposals is mixed in the full Senate and House (see Freeman, 1983; Maraniss, 1984), the ultimate form such legislation takes depends on resolving regional disagreement not only about the nature of the problem the timing of any program but also how to allocate the costs of control.

SCIENCE AND ECONOMICS

Robert Angus Smith, an English chemist, might well claim title of being the "father of acid rain." Smith's pioneering studies of precipitation chemistry and its effects first used the term "acid rain." Drawing upon data from England, Scotland and Germany, he demonstrated that variation in regional factors such as coal combustion, wind trajectories, the amount of frequency of precipitation, proximity to seacoasts, and the decomposition of organic materials affected sulfate concentrations in precipitation (Smith, 1872). Smith's work, however, was largely ignored and failed to generate follow-up research.

Contemporary concern originated in three seemingly unrelated areas: limnology, agricultural science, and atmospheric chemistry. Oden (1968), in the first major attempt to integrate knowledge from those disciplines, maintained that air mass trajectories, when matched to precipitation chemistry data, indicated that sulfur and nitrogen were transported long distances (100-2000 km) and that clearly identifiable source and receptor areas existed. His analysis formed the initial basis for concluding that acid deposition is a large-scale regional phenomenon with adverse ecological consequences. Thus, information produced by a complex and rapidly evolving body of research helps to define the acid deposition problem.

Unfortunately, the simplicity of the term "acid rain" conveys the image of an easily measured and well understood phenomenon. In fact, the problem stems from a series of complex and varied chemical, meteorological, and physical interactions (Regens,

FIGURE 4.1
Annual mean value of pH in precipitation[1]

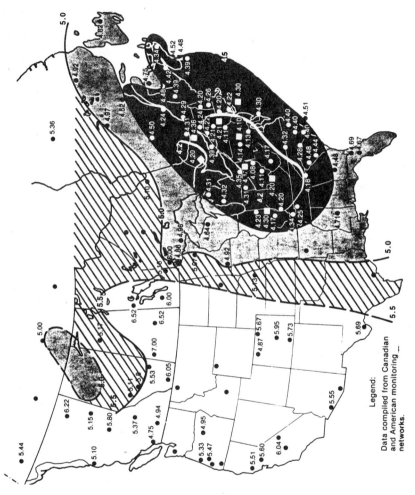

[1]Weighted by the amount of precipitation in the United States and Canada.

Source: U.S.-Canada Work Group 2. Atmospheric Science and Analysis, Final Report. Washington, D.C.: U.S. EPA, 1982.

1984). For example, monitoring programs have focused on measuring only the precipitation (wet) component. Values for total (wet + dry) deposition consist of estimates because valid and reliable methods for directly measuring the dry component on a field network basis are still undergoing development.

With this limitation in mind, precipitation chemistry data collected in the 1970s indicate that acid deposition occurs throughout eastern North America. The area of greatest acidity is concentrated over eastern Ohio, western New York, and northern West Virginia (U.S.-Canada Work Group 2, 1982). Lewis and Grant (1980) found that higher elevation, rural areas in the western United States also have low pH rainfall probably due to significant increases in nitrogen oxide (NO_x) emissions throughout the region. While broad geographical generalizations derived from individual monitoring site data are subject to considerable uncertainties,[6] Figure 4.1 suggests the potential for near continential-scale impacts if adverse effects are linked to inputs (loadings) of acidic compounds above a given level or rate.

For eastern North America, SO_2 and NO_x emissions from man-made sources are estimated to exceed those from natural sources such as ocean-land fluxes and vegetation by at least a factor of ten (U.S. Environmental Protection Agency, 1983). As a result, because the pH of rainfall would seem to be more acidic than one might expect due to natural processes, the monitoring data encourage policymakers to seek information about the geographical distribution of man-made sources of precursor emissions. Figure 4.2 reveals that the areas receiving the greatest amounts of acid deposition on an annual average basis are found within or downwind from and relatively proximate to major man-made emissions source regions. This indicates that man-made (i.e., anthropogenic) sources contribute overwhelmingly to the emissions of acidic precursors.

Clearly, subject to technological and economic constraints, emissions from man-made sources--such as electric utilities, industrial boilers, and motor vehicles--can be limited through fuel switching, pre-combustion cleaning and/or post-combustion flue gas desulfurization (FGD). As a result, assumptions about the relationship between emission sources and receptor areas form the basis for formulating an acid rain control strategy. As we noted earlier, $SO_4^=$ is the major constituent of acid rain in eastern North America as well as Europe. This makes it possible to outline at least some of the initial elements of a reduction strategy. Moreover, because public policies represent a response to perceived problems, this raises the

FIGURE 4.2
Sulfur dioxide and nitrogen oxides emissions[1]

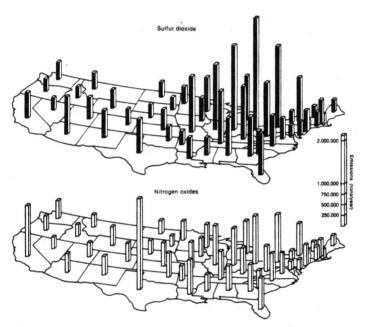

[1]Representative values for 1980.

Source: U.S.-Canada Work Group 3B. Emissions, Costs and Engineering Assessment, Final Report. Washington, D.C.: U.S. EPA, 1982. (And: G. Gschwandtner, et al., "Historic Emissions of Sulfur and Nitrogen Oxides in the United States from 1900 to 1980," draft report to EPA, 1983.)

question of acid rain's environmental effects. Conclusive evidence exists for chemical and biological changes, including fish kills, in lakes and streams which have limited capacities to neutralize acidic inputs. Evidence of damages to non-aquatic ecosystems, especially forests, is largely circumstantial. Yet impacts are plausible and evidence of effects is growing. For example, adverse effects on forests may result from the leaching of nonagricultural soil nutrients or through the mobilization of toxic metals. Growing concern also has been expressed about the impact of acid deposition on cultural artifacts, buildings and other structures. Field studies have linked such damage to air pollution, especially in urban areas with high ambient SO_2 concentrations. Unlike respirable sulfates or fine particulate matter, however, acid deposition does not appear to represent a direct risk to human health; however, limited health risks may be associated with episodic acid fog events or the leaching of metals such as lead into drinking water supplies.

Crocker and Regens (1985) examine the potential as well as limits of formal economic analyses to guide policy choice in the acid rain area. They note that a limited number of attempts have been made to assess those benefits and costs.[7] Deriving benefit-cost estimates depends upon the analyst making a variety of assumptions. Evidently uncertainty surrounds both benefits and costs. Nonetheless, information about control costs is somewhat more tangible at least in an aggregate sense.

A number of studies provide information about the aggregate costs for reducing the precursors of acid deposition (Rubin, 1983; Office of Technology Assessment, 1982; U.S.-Canada Work Group 3B, 1982; McGlamery and Torstick, 1976). Those analyses provide substantial insight into the costs under alternative strategies, especially for capturing SO_2 reductions. Annualized cost estimates of achieving aggregate SO_2 reductions in the utility sector have been quite uniform across studies. Annualized costs range from $1 - $2 billion for a 40% reduction, from $2 - $4 billion for a 50% reduction, and from $5 - $6 billion for 66-75% reductions.[8] The estimated average increases in electricity rates to accomplish such a reduction in utility industry SO_2 emissions have ranged from 1.4% for a 4 million ton rollback to 8.0% for a 12 million ton rollback. Naturally, rate increases as well as control costs for individual utility systems may be substantially higher than the average values.

Alternative interpretations of the science and economics of acid rain are plausible. Existing information can be used to construct a policy rationale for either maintaining the status quo or taking further action now with respect to initiating acid deposition control measures. This is true for several reasons. First, although its extent as well as the rate at which damages are induced remains uncertain, the perception that acid deposition is an environmental problem combined with its salience on the policy-setting agenda compels a response by government. Second, man-made sources generally are the overwhelming contributors to acid deposition in eastern North America. While expensive, control technology is available to reduce significantly emissions from those sources (see OECD, 1983b). Third, a recent National Academy of Sciences report concludes that it is reasonable to assume that reductions in SO_2 emissions over a broad area for several years will produce an essentially proportionate reduction in annual average $SO_4^=$ deposition for that area (National Research Council, 1983). Finally, other parameters of air quality in eastern North America-- regional visibility, particulate matter loadings, and ambient SO_2 levels--are affected strongly by the precursors to acid deposition, and they are likely to be improved if atmospheric loadings of precursor emissions are reduced. Thus, while uncertainties about nonaquatic ecosystem effects and site-specific changes in deposition patterns and pH within sensitive receptor areas persist, it is feasible to outline the elements of a control program focusing on SO_2 reductions. As a consequence, it is important to consider the revenue generation alternatives available for financing such a program.

FINANCING CONTROL COSTS

Acid rain forces policymakers to confront a typical environmental controversy. While a scientific foundation for some kind of government action exists, popular perceptions of environmental risk may be more powerful motivations for government intervention than actual scientific evidence. Moreover, the United States has a long tradition of undertaking regulatory action to demonstrate public concern or simply to "do something." And, as a society, we often implement risk-reduction strategies on the basis of only fragmentary evidence of hazards themselves or the relative costs and benefits of alternative strategies (see Crandall and Lave, 1981). Senator John Glenn (D-Ohio) has noted: "The crux of the acid rain cleanup problem has always been the cost of cleanup and who should bear it" (Mosher, 1983, p. 998). As a result, the financing issue, including questions of equity, increasingly has become the focus of both the public policy debate and the symbolic politics surrounding the issue of allocating the costs of SO_2 reduction program.

In one of the more commonly discussed financial alternatives, the Congress or the Environmental Protection Agency (EPA) would set emission targets. Each state would then allocate its share of the required reductions among the electric utilities in that state. Two scenarios exist for allowing the individual utility systems to determine how to achieve their reduction quotas. Under one scenario, each utility would decide on an appropriate strategy to achieve the necessary emissions reduction. The utility would then ask its state public utility commission (PUC) to permit the level of rate increases necessary to offset the additional control costs. This scenario leaves the actual choice of specific compliance approach (i.e., scrubbing via FGD retrofit, fuel switching, physical coal cleaning) to the discretion of the utility and/or the various state governments. The second scenario also would use the rate-making process to generate capital for financing acid rain controls, but the control options would be statutorily mandated.

Because a program for reducing SO_2 emissions diverts capital away from availability for investment in other sectors of the economy, the electric utility rate-making process is one of the more economically efficient mechanisms available for funding such a program.[9] The incremental administrative costs should be relatively low, especially when compared with other approaches since the rate-making process relies on an established system. In addition, reliance on the rate-making process results in imposing control costs on those utilities which are required to reduce emissions consistent with the "polluter pays" principle which underlies most existing U.S. environmental statutes.

Such an approach, however, is not without some disadvantages. Reliance on the utility rate-making process to finance acid deposition controls would tend to concentrate the costs of SO_2 emissions reductions on utilities burning high-sulfur coal. Table 4.1 indicates that electricity consumers in some states would receive relatively large rate increases in the initial years of such a control program. For example, a 10 million ton reduction might increase utility rate in the Midwest an average of 7.4 percent. While those midwestern states could partially or totally protect their low-income consumers from absorbing such an increase with some form of assistance, such a financial subsidy would require allocating a greater share of the overall control costs to more affluent individuals and/or industrial customers in those states adopting an assistance program. Finally, the rate-making system tends to "front-load" control costs into the early

TABLE 4.1
Percent change in electricity rates for
10 million ton reduction[a]

Region	State	Percent Increase[b]
New England	Connecticut	6.1
	Massachusetts	6.1
	Maine	3.2
	New Hampshire	3.2
	Rhode Island	6.1
	Vermont	3.2
Average		4.7
Mid-Atlantic	Delaware	3.2
	Maryland	3.2
	New Jersey	2.3
	New York	2.8
	Pennsylvania	6.5
Average		
Midwest	Iowa	3.3
	Illinois	0.7
	Indiana	13.5
	Michigan	3.6
	Minnesota	-0.9
	Missouri	13.0
	Ohio	17.8
	Wisconsin	8.3
Average		7.4
South	Alabama	-2.1
	Arkansas	2.9
	Florida	3.7
	Georgia	4.6
	Kentucky	10.8
	Louisiana	0.4
	Mississippi	12.8
	North Carolina	1.2
	South Carolina	1.2
	Tennessee	10.1
	Virginia	2.6
	West Virginia	6.0
Average		4.5

[a] Costs based on % charge in electricity rates in 1990 for a 1st year revenue requirement on a composite bill assuming intrastate trading.

[b] Excludes the District of Columbia.
Source: Adapted from Wetstone (1983).

years of the payback period. This feature could make the first year's rates as much as 50% higher than the long-term average costs of the program. Ratemaking reforms, such as allowing charges for Construction Work in Progress (CWIP), could equalize rate increases over time. However, because the normal rate-making process forces utilities to raise the funds for financing acid rain controls through the capital market, utility systems with low allowable rates of return, poor prospects for future growth, and/or high indebtrnedness are not likely to be capable of financing controls without some public sector assistance. As a consequence, irrespective of the science of acid rain, unless some spread-the-cost financing scheme is adopted, utilities are likely to oppose control proposals (Magnet, 1983).

Alternatives to the normal electric utility rate-making process which might, at least partially, shift the burden of control costs are generally less economically efficient. The alternatives involve raising all or part of the funds for financing controls on the basis of electricity production, emissions, fuel use, general revenues, or a combination of these sources.[10] Those options may be preferred for several reasons. First, Table 4.2 reveals that most alternatives allocate the costs of financing an SO_2 emissions reduction program over a broader segment of the population. Expansion of the base thereby reduces the probability that any one group receives substantial electricity rate hikes.[11] The options also provide opportunities to target funds on the basis of additional social or economic goals. For example, revenues could be used to provide subsidies to severely impacted ratepayers, especially low income consumers. They also might be used to mitigate adverse employment impact in the Midwest due to shifts in the level of high-sulfur coal production.

IMPLICATIONS FOR POLICYMAKING

The thrust of our earlier summary of the science and economics of acid rain, coupled with the fact that each of the financing alternatives can be set at a level sufficient to have essentially equivalent revenue generation capability relative to the other options, force policymakers to confront the reality that the selection of an option ultimately becomes a function of distributional considerations. Howard and Perley (1980: 185-186) succinctly state the obvious conclusion one must draw: "Acid rain is an environmental crisis with sweeping financial implications, but its solution is political." In other

TABLE 4.2
Distributional impact of selected options
for financing acid deposition reductions
(revenue percentage/state)

State	Generation Fee[a]	Btu Tax[b]	Emissions Fee[c]
Alabama	2.56	2.02	2.86
Alaska	0.13	0.41	0.07
Arizona	1.92	1.06	3.39
Arkansas	0.85	0.92	0.38
California	5.51	7.98	1.68
Colorado	1.32	1.24	0.50
Connecticut	0.62	0.88	0.27
Delaware	0.47	0.34	0.41
District of Columbia	0.02	0.11	0.06
Florida	4.78	3.30	4.12
Georgia	2.98	2.27	3.16
Hawaii	0.35	0.37	0.22
Idaho	--[d]	0.27	0.18
Illinois	3.93	4.96	5.54
Indiana	3.82	3.46	7.56
Iowa	1.10	1.30	1.24
Kansas	1.36	1.52	0.84
Kentucky	3.27	1.91	4.22
Louisiana	2.51	4.51	1.14
Maine	0.12	0.32	0.36
Maryland	0.97	1.36	1.27
Massachusetts	1.60	1.63	1.30
Michigan	3.25	3.64	3.41
Minnesota	1.16	1.54	0.98
Mississippi	0.84	1.04	1.07
Missouri	2.73	2.07	4.90
Montana	0.30	0.43	0.62
Nebraska	0.50	0.68	0.28
Nevada	0.80	0.44	0.91
New Hampshire	0.25	0.26	0.35
New Jersey	1.15	2.45	1.05
New Mexico	1.30	0.92	1.01
New York	3.56	4.71	3.56
North Carolina	3.57	2.16	2.27
North Dakota	0.78	0.48	0.40
Ohio	6.10	5.53	9.97
Oklahoma	2.44	1.85	0.45
Oregon	0.10	0.66	0.23
Pennsylvania	5.78	5.53	7.61
Rhode Island	0.05	0.20	0.06
South Carolina	1.29	1.10	1.23
South Dakota	0.15	0.26	0.15
Tennessee	2.75	1.91	4.05
Texas	11.71	12.10	4.81
Utah	0.63	0.71	0.27
Vermont	--[d]	0.10	0.02
Virginia	1.10	1.67	1.36
Washington	0.40	1.16	1.02
West Virginia	4.16	1.74	4.10
Wisconsin	1.43	1.71	2.40
Wyoming	1.45	0.78	0.69

[a] Based on 1981 conventional steam generation (i.e., non-hydro, non-nuclear) of electricity expressed as 10^6 kwh. See Edison Electric Institute (1983).

[b] Based on 1980 consumption of fossil fuels (coal, natural gas, and petroleum products) expressed as 10^{12} Btu. See U.S. Department of Energy (1983).

[c] Based on 1980 SO_2 emissions estimates expressed as 10^6 tonnes. See U.S.-Canada Work Group 3B (1982).

[d] State shares < .001 percent of total U.S. revenue requirement.

words, classic equity issues--who gains the benefits and who bears the burden of the costs and risks--are likely to drive the decision to embark upon any particular path. Rhodes and Middleton (1983, p. 7) have made this point as follows:

> As acid rain control measures are proposed and debated, an important observation must be taken into account in the formulation of regulatory policy; namely, that the current unregulated acid rain situation produces <u>certain benefits</u> to some and <u>uncertain costs</u> to others. Proposed regulation of acid-causing emissions, however, promises <u>certain costs</u> for those who presently enjoy the certain benefits but <u>uncertain benefits</u> for those who presently incur the uncertain costs.

In such circumstances, the economic efficiency of any financial strategy may be of only marginal importance. The very different patterns of bureaucratic politics and interest-group mobilization and interaction, including the way in which these key participants use information from the physical and natural sciences or economics, are probably more significant in determining the ultimate resolution of the acid rain debate.

Acid rain provides another striking example of the salience of politics and the unpredictability of bureaucratic strategies and interest-group coalitions, in part, because concerns about acid deposition are intimately linked to the larger issue of air quality management. In fact, for years the acid rain issue tended to be viewed by all participants as a secondary component of the Clean Air Act implementation controversy. Even when discussed, the acid rain issue was viewed as a subset of broader energy/environment concerns.

With action by the EPA apparently stalemated for at least the short-term, and with other major bureaucratic actors--such as the Department of Interior and Department of Energy--captured by the same producer interests (see Tolchin and Tolchin, 1983), the focus of attention has shifted to the Congress. However, most of the debate to date has downplayed the ways in which various proposals might shape interest-group politics in the future. Yet the very different patterns of actor mobilization and interaction which are likely to occur as a function of the allocation of acid rain control benefits and costs appear to be central to the ongoing policy debate.

Like most environmental issues, the acid rain problem does not fit easily into a framework characterized by concentrated costs <u>and</u> benefits. We know substantially less about the benefit side of the acid deposition equation (i.e., value of damages prevented or mitigated) than we do about costs. But we can anticipate that benefits will accrue to different parties-at-interest, at different points in time, and at different magnitudes than will be the case with costs (Regens and Crocker, 1984). And we also know that a strong case can be made that the U.S. political system and the international system function best when the various parties-at-interest do have an understanding of distributional consequences (i.e., how they stand to benefit from or pay for a particular policy). Environmental policy-making is no exception to this rule (Rosenbaum 1977: 21). Acid rain control decisions ultimately will be resolved according to precisely this kind of distributive calculus since they represent an explicit political choice. In a democracy, such public choise, to be legitimate should be made within the political rather than private market context.

NOTES

1. "Acid rain" refers to what is more precisely identified as the wet and dry processes for the deposition of acidic inputs to ecosystems. Acidity is measured on the logarithmic pH scale (pH equals the negative $\log_{10}$ of the hydrogen ion concentration); a solution that is neutral has pH 7.0. The "natural" acidity value often is assumed to be pH 5.6 calculated for distilled H_2O in equilibrium with atmospheric CO_2 concentrations but the presence of other naturally occurring species--SO_2, NH_3, organic compounds, windblown dust--can produce "natural" values of pH 4.9 to 6.5 (Charlson and Rodhe, 1982; Galloway et. al., 1982).
2. Austria, Denmark, the Federal Republic of Germany, Finland, France, the Netherlands, Norway, Sweden, and Switzerland.
3. The EB is composed of national representatives of all countries who have ratified the ECE Convention and is responsible for developing protocols for implementing the terms of the Convention. Belgium, Luxembourg, Lichtenstein, Italy, and the Soviet Union, the German Democratic Republic, Bulgaria, and Czechoslovakia joined the original "30% Club" in endorsing the reduction.
4. In 1980, annual SO_2 emissions from utility sources in the 31-state region totalled approximately 17 million tons (see U.S.-Canada Work Group 3B, 1982).
5. The top 50 plants located in the following states-- Alabama (1); Florida (3); Georgia (2);

Illinois (4); Indiana (7); Kentucky (4); Michigan (1); Missouri (5); Ohio (8); Pennsylvania (4); Tennessee (5); West Virginia (4); and Wisconsin (1)--had the greatest tonnage of SO_2 emissions from among the set of power plants emitting SO_2 at a rate of $\geq$ 3 lb/mmBtu (see Office of Technology Assessment, 1983).

6. Work Group 2 of the U.S.-Canada MOI (1983) concluded that individual monitoring station values were reasonably accurate but that the uncertainty in isopleth map lines generalized from site data was about $\pm$20 percent in magnitude and 50-200 km in position for eastern North America. The degree of uncertainty would be higher for western North America because fewer monitoring stations are located there.

7. Economic analyses generally do no more than multiply natural science findings about service flow changes by an invariant price. The researchers then speculate, if they recognize them at all, how the resulting estimates would differ if price responses and agent adaptations were captured. Because of the differences in the behavior of emitters and receptors when a market in emissions rights does and does not exist, and because of the lack of parallel markets, economically efficient outcomes for the acid deposition problem may be impossible to trace exhaustively. The economic criterion is then reduced to whether those who gain from a change in precursor control could, in principle, compensate the losers and still have some residual gain. For a discussion of the theoretical basis for applying the technique to acid deposition control see Regens and Crocker (1984).

8. None of the cost studies explicitly consider the rather unique decision problems of the utility industry, the manner in which these problems influence its institutionalized habits and modes of thinking, and therefore the observed costs on which the aforementioned studies are founded. There are circumstances under which costs of control in the industry are likely to be above those that would be experienced by profit-maximizing, perfectly competitive producers of the identical type of output (Goldberg, 1976; Averch and Johnson, 1962). The size of the increase is unknown.

9. An efficiently implemented eight million ton regional roll-back in SO_2 emissions is estimated to cost $3.7 billion, on an annualized basis, in 1995. Utilities would pay $3.1 billion and other industries would pay an additional $.6 billion. These cost estimates are based on ICF's analysis of S. 3041 (see ICF, Inc., 1983).

10. All of these options would give two more Congressional committees jurisdiction over legislation for acid rain control--the House Ways and Means and the

Senate Finance Committees.
11. An argument can be made that it is less disruptive and, therfore, more socially desirable to require a number of small adjustments rather than a few large ones.

REFERENCES

Altshuller, A. P., and G. A. McBean. **The LRTAP Problem in North America: A Preliminary Overview Prepared by the United States-Canada Bilateral Research Consultation Group on the Long-Range Transport of Air Pollutants.** Downsview, Ontario, Canada: Atmospheric Environmental Service, 1979.

Averch, H., and L. L. Johnson. "Behavior of the Firm under Regulatory Constraint." **American Ecnomic Review.** 52 (December 1962). 1052-1069.

Charlson, R. J., and H. Rodhe. "Factors Controlling the Acidity of Natural Rainwater." **Nature.** 295 (1982). 683-685.

Cowling, E.B. "Acid Precipitation in Historical Perspective." **Environmental Science & Technology.** 16 (February 1982). 110A-123A.

Cowling, E. B. **An Historical Resume of Progress in Scientific and Public Understanding of Acid Precipitation and Its Consequences.** Oslo-As, Norway: SNSF Project, 1981.

Crandall, R. W., and L. B. Lave (eds.). **The Scientific Basis of Health and Safety Regulation.** Washington: The Brookings Institution, 1981.

Crocker, T.D., and J. L. Regens. "Benefit-Cost Analyses of Acid Deposition Control." **Environmental Science & Technology.** 19 (February 1985). 112-116.

Edison Electric Institute. **Statistical Yearbook of the Electric Utility Industry/1982.** Washington: Edison Electric Institute, 1983.

Editorial Research Reports. **Environmental Issues: Prospects and Problems.** Washington, D.C.: Congressional Quarterly, Inc., 1982.

Freeman, G. C., Jr. "The Politics of Acid Rain" (presented at the fall meeting of the Board of Directors of the Southeastern Electric Exchange). Hilton Head, S.C. September 1983.

Galloway, J.N.; G. E. Likens; W. C. Keene; and J. M. Miller. "The Composition of Precipitation in Remote Areas of the World." **Journal of Geophysical Research.** 1982. 8771-8786.

Goldberg, V. P. "Regulation and Administered Contracts." **Bell Journal of Economics.** 7 (Autumn 1976). 426-448.

Howard, R., and M. Perley. **Acid Rain: Its Devastating Impact on North America.** New York: McGraw-Hill, 1980.

ICF, Inc. **Analysis of a Senate Emission Reduction Bill.** Washington: ICF, Inc., February 1983.

MacNeill, J. M. "Coal and Environment: Constraint or an Opportunity." In E. S. Rubin and I. M. Torrens, eds., **Costs of Coal Pollution Abatement.** Paris, France: Organization for Ecnomic Cooperation and Development, 1983. 55-62.

Magnet, M. "How Acid Rain Might Dampen the Utilities." **Fortune.** 108 (August 8, 1983. 57-64.

Maraniss, D. "Congress' Search for an Acid Rainbow." **The Washington Post National Weekly Edition.** 1 (February 13, 1984). 6-7.

McGlamery, G. G., and R. L. Torstick. "Cost Comparisons of Flue Gas Desulfurization Systems." In K. E. Noll and W. T. Davis, eds., **Power Generation: Air Pollution Monitoring and Control.** Ann Arbor, MI: Ann Arbor Science Publishers, 1976.

Mosher, L. "Acid Rain Debate May Play a Role in the 1984 Presidential Sweepstakes." **National Journal.** 15 (May 14, 1983). 1998-999.

National Research Council. **Acid Deposition: Atmospheric Processes in Eastern North America: A Review of Current Scientific Understanding.** Washington, D.C.: National Academy Press, 1983.

Oden, S. "The Acidification of Air and Precipitation and Its Consequences in the Natural Environment." **Ecology Committee Bulletin Number 1.** Swedish National Science Research Council, 1968.

Office of Technology Assessment. "An Analysis of the Sikorski/Waxman Acid Rain Control Proposal: H.R. 3400, 'The National Acid Deposition Control Act of 1983'." Washington, D.C.: U.S. Congress, Office of Technology Assessment, July 1983.

Office of Technology Assessment. **The Regional Implications of Transported Air Pollutants: An Assessment of Acidic Deposition and Ozone** [Interim Draft]. Washington, D.C.: U.S. Congress, Office of Technology Assessment, July 1982.

Organization for Economic Cooperation and Development. **Coal: Environmental Issues and Remedies.** Paris, France: Organization for Economic Cooperation and Development, 1983a.

Organization for Economic Cooperation and Development. **Coal and Environmental Protection: Costs and Costing Methods.** Paris, France: Organization for

Economic Cooperation and Development, 1983b.
Regens, J. L. "Acid Rain: Does Science Dictate Policy or Policy Dictate Science?" In T. D. Crocker, ed., **Economic Perspectives on Acid Deposition.** Woburn, MA: Butterworth Scientific Publishers, 1984. 5-19.
Regens, J. L. and T. D. Crocker. "Applying Benefit-Cost Analysis to Acid Rain Control." **Management Science and Policy Analysis.** 1 (Winter 1984). 12-17.
Rhodes, S. L., and P. Middleton. "The Complex Challenge of Controlling Acid Rain." **Environment.** 25 (May 1983). 6-9; 31-37.
Royal Ministry for Foreign Affairs, Royal Ministry of Agriculture. **Air Pollution Across National Boundaries: The Impact on the Environment of Sulfur in Air and Precipitation.** Sweden's Case Study for the United Nations Conference on the Human Environment, Stockholm, Sweden, 1972.
Rosenbaum, W. A. **The Politics of Environmental Concern.** (2d ed.) New York: Praeger, 1977.
Rubin, E. S. "International Pollution Control Costs of Coal-Fired Power Plants." **Environmental Science and Technology.** 17 (August 1983). 366A-377A.
Smith, R. A. **Air and Rain: The Beginnings of Chemical Climatology.** London: Longmans, Green, 1972.
Tolchin, S. J., and M. Tolchin. **Dismantling America: The Rush to Deregulate.** Boston: Houghton Mifflin, 1983.
Ulrich, B. **Immissionsbelastungen Von Waldokosystemen** [Dangers to the Forest Ecosystem Due to Acid Precipitation]. Special Communication from: the Landesanstalt fur Okologie Landschaftsentwicklung und Forstplanung Nordrhein-Westfalen, Federal Republic of Germany, 1982.
U.S.-Canada Work Group 2. **Atmospheric Science and Analysis, Final Report.** Washington: U.S. Environmental Protection Agency, 1982.
U.S.-Canada Work Group 3B. **Emissions, Costs and Engineering Assessment, Final Report.** Washington: U.S. Environmental Protection Agency, 1982.
U.S. Department of Energy. **State Energy Data Report.** Washington: U.S. Department of Energy, 1983.
U.S. Environmental Protection Agency. **The Acidic Deposition Phenomenon and Its Effects: Critical Assessment Review Papers, Public Review Draft** [Vol. I]. Washington: U.S. Environmental Protection Agency, 1983.
Vermeulen, A. J. "Acid Precipitation in the Netherlands." **Environmental Science & Technology.** 12 (1978).
Wetstone, G. S. "Paying for Acid Rain Control: An

Introduction to the Trust Fund Approach." **The Environmental Forum.** 2 (August 1983). 14-20.

Wright, R. F., R. Harriman, A. Henriksen, B. Morrison, and L. A. Caines. "Acid Lakes and Streams in the Galloway Area, Southwestern Scotland." In D. Drablos and A. Tollan, eds., Ecological Impact of Acid Precipitation. Oslo-As, Norway: SNSF Project, 1980.

5
Acid Rain Legislation and the Clean Air Act: Time to Raise the Bridge or Lower the River?

Larry B. Parker
and John E. Blodgett

INTRODUCTION

During recent congressional sessions, the topic of acid precipitation and its control has dominated debates about air quality issues. Surprisingly, many participants in these legislative debates seem to view acid rain as an issue-area separate from other air quality matters. Legislatively, this separation is clearly an artificial one, as illustrated by the treatment of acid rain legislation by the Senate Environment and Public Works Committee as one part of comprehensive amendments to the Clean Air Act (CAA). A few commentators admittedly have analyzed the extent to which proposed acid precipitation control programs would fit into existing CAA programs.[1] Nevertheless, much of the debate has been conducted without continuing reference to the Clean Air Act and, as a result, several important issues have been obscured.

These issues include: Does the present impasse over the appropriateness of regulatory controls to mitigate acid precipitation reflect a lack of administrative will, a defect in the CAA as written, or a genuine inadequacy of scientific understanding of the nature of the problem and of how to control it--or a mix of all three? Would the proposals being most actively debated in the Congress complement or contradict existing CAA authorities? Given present air quality program, how would the acid precipitation problem be affected by possible future socio-economic trends affecting precursor emissions?

The 1970 CAA was based on contemporaneous perceptions of present and future economic and environmental conditions. Of particular import to the present acid precipitation issue, those perceptions included the expectation that rapid energy growth--even with stringent controls imposed on new fossil fuel fired boilers--would result in SO_2 emissions levels remaining constant at best and potentially increasing, albeit at a slower rate than without controls. At the same time, the most pressing concern of legislators supporting the Clean Air Act was their desire to curb the human health impact of air pollution. Since 1970, however, these parameters have changed. Energy growth has declined substantially, and SO_2 emissions are actually lower. Still, with public and environmentalist concern shifting increasingly to "welfare" values (such as forest, fish, and visibility), even current levels of SO_2 emissions may be causing demonstrable harm. Although the 1970 CAA includes provisions to protect such values, in the eyes of many they are proving ineffective in coping with effects of acid precipitation.

This chapter analyzes the present status of the CAA vis-a-vis the acid precipitation problem. Like most of the major acid precipitation bills before Congress, it focuses on SO_2 emissions from utilities as representative of the overall problem. It begins with a review of what has happened from 1970 to the present in order to assess how the CAA has affected SO_2 emissions and how its authorities relate to the acid precipitation problem. The analysis then turns to how those present authorities of the CAA might address acid precipitation in the future in the light of currently projected energy-economic trends. The findings have implications for present economic or energy-related policies that might affect likely time-frames over which SO_2 emissions can be expected to fluctuate, and for environmental policies that would treat acid precipitation as a unique, circumscribable problem as opposed to a generic problem of environmental degradation.

The acid precipitation problem has been frequently and extensively described elsewhere. For the purposes of this analysis, the reader need only recognize that four characteristics of the problem are crucial: (1) the known adverse effects of acid precipitation concern not human health, but "welfare" values; (2) the chemical compounds comprising acid rain and proximately responsible for the adverse effects are not the pollutants emitted by various sources, but are instead the transformed products of those emissions; (3) the source of emissions and the location where the transformed products are deposited may be separated by up to hundreds of miles; and (4) at any given time, the total amount of acid precipitation is likely to be more important to causing adverse environmental effects than are the concentrations of precursor pollutants.

THE CLEAN AIR ACT AND ACID RAIN

As enacted in 1970 the Clean Air Act establishes a two-pronged approach to the protection and enhancement of the quality of the Nation's air. First, the Act establishes National Ambient Air Quality Standards (NAAQS), which set limits on the levels of air pollutants in ambient air. Second, the Act requires national, technology-based emission limits on new sources.

National Ambient Air Quality Standards

The NAAQS are designed to prevent emissions of specific air pollutants from rising to levels causing adverse effects on the public health or welfare. For each air pollutant of concern, EPA prepares a "criteria document," which lays out the best scientific information on the relationship between levels of the pollutant and adverse effects. Based on this document, EPA then sets the NAAQS necessary to prevent adverse effects. Each State is responsible for developing a plan (State Implementation Plan, of SIP) for achieving and maintaining NAAQS. The NAAQS itself does not directly concern regulatory controls on specific sources; it is the SIP that must provide for emissions reductions adequate to achieve and maintain NAAQS.

The CAA provides for two levels of NAAQS: primary standards, set at a level to protect the public health, with an "adequate margin of safety"; and secondary standards, set at a level to protect the public welfare. The Act defines the public welfare as including "effects on soils, water, crops, vegetation, man-made materials, animals, wildlife, weather, visibility, and climate, damage to and deterioration of property, and hazards to transportation, as well as effects on economic values and on personal comfort and well-being" [sec. 302(h)]. The Act sets specific deadlines for the attainment of the primary NAAQS, but not for the secondary ones.

Except for continuing problems in certain urban areas, this strategy of controlling air pollution has been generally successful in keeping pollutant concentration below levels believed to pose a health threat. But where even lower concentrations may pose a threat to "welfare" values--such as the fish and forests and visibility affected or potentially affected--by acid precipitation, these procedures of the CAA have not been effective. One reason why the NAAQS have not effectively addressed the acid precipitation problem concerns the weight of evidence of the problem. Other reasons for this failure concern difficulties in relating CAA procedures to characteristics unique to the acid precipitation issue.

To amplify on the first reason, in the eyes of many key decision-makers, the scientific information on the acidity of precipitation, on specific, measurable adverse effects, and on the relationships between them remains insufficient to trigger the process of setting the secondary NAAQS at a level adequate to protect against those effects and then of compelling the revision of SIPs to achieve them. But even if the process were triggered, certain difficulties lead one to suspect that the process still might not be particularly effective, even though some difficulties were recognized and addressed in 1977 amendments to the Act.

Four of these difficulties are as follows:

--Acid precipitation concerns secondary adverse impacts, and the Act does not establish statutory deadlines for attaining the secondary NAAQS. At present, the Act requires that a SIP specify a "reasonable time" at which any secondary standard shall be attained.

--Acid precipitation results at least in part from long-distant transport of pollutants. The SIP process basically presumes that air quality is protected if each State ensures that NAAQS can be and will be maintained within each State by that State controlling as necessary sources within the State. The recognition that an upwind State's sources could contribute to nonattainment (sec. 126) in the CAA to deal with interstate air pollution. However, its procedures become increasingly difficult to engage as the distance between the source and the deposition increases and the source/cause and receptor/effect relationship becomes less direct.

--The acid precipitation pollutants--primarily sulfates (SO_4^-) and nitrates (NO_4) causing adverse effects at receptor sites-- are not in the same chemical form as the emitted pollutants--primarily sulfur dioxide (SO_2) and certain nitrogen oxides (NO_x). This transformation of pollutants means that the SO_2 and NO_x NAAQS do not directly address acid precipitation. The 1977 Amendments to the CAA directed EPA to investigate how sulfates are formed and how to protect the public health and welfare from any adverse effects. These studies are now in process. Since the Amendments do not set a deadline for completion of the studies and since they have become embroiled in the acid precipitation debate, it is not clear when they will be completed.

--Finally, but perhaps of first importance, the NAAQS address concentrations of pollution, but not total loadings. That is, as long as the maximum allowable concentration of a pollutant is not exceeded, the total amount of the pollutant in the air does not trigger regulatory attention. The impacts of acid precipitaiton, however, appear to result from total loadings of acid-causing pollutants. As a result, those proposing controls to prevent acid precipitation impacts would measure the maximum pollution level in terms of total deposition--for example, a maximum of 18 pounds per acre of sulfate--rather than in terms of a maximum concentration in ambient air.

On this last issue, the 1970 Clean Air Act Amendments had the effect of exacerbating this problem by not prohibiting facilities from dispersing their pollutants. That is, if the emissions from a facility led to concentrations of a pollutant in excess of a NAAQS and were therefore subject to abatement under a SIP, the facility, instead of reducing emissions so the NAAQS wouldn't be violated, could raise its smokestacks so as to dilute the ambient concentration of the emissions. Recognizing this problem, Congress in the 1977 Amendments to the CAA prohibited the use of dispersion techniques as a way of meeting air quality requirements (sec. 123). Nevertheless, the problem remains that SO_2 emissions not exceeding present NAAQS could contribute to acid precipitation that would cumulatively reach levels having the potential to cause adverse effects.

Despite these problems, several proposals for using particular existing CAA authorities to address the acid precipitation problem have been advanced. Following a suggestion by the Natural Resources Defense Council (NRDC) that tighter EPA compliance monitoring and enforcement of SO_2 emissions would reduce acid precipitation-causing SO_2 emissions by two million tons per year, the **Environment Reporter** prepared a special report.[2] The NRDC recommendation focused on a number of issues--including EPA's allowance of some States to relax SO_2 emission limitations under SIPs; the averaging time which EPA has proposed for assessing compliance; EPA's method of monitoring emissions; and EPA's tall stacks policy. The article goes on to review the arguments, including the positions of EPA, the industry, and State officials.

A new set of proposals for using existing CAA provisions is being explored by Eugene M. Trisko, a Washington attorney and consultant on energy/environment issues. Trisko suggests that EPA has a nondiscretionary duty to set an ambient sulfate standard and that acid precipitation control could be combined with CAA authority to protect visibility under section 169A.[3] Finally, he advances the possibility of controlling SO_2 through an emissions density zoning concept as a way of making reduction requirements correlate more closely with deposition levels. In his opinion, the emissions density zoning concept combined with reliance on existing CAA authorities could reduce the Congressional controversies and facilitate reauthorization of the Act with an acid precipitation abatement program. In his words:

> Using well-established Clean Air Act mechanisms to combat acid rain would liberate Congress form the paralysis which has immobilized reauthorization of the act since 1982.... Since it is apparent that acid rain is simply a catch phrase for sulfate control, it seems timely for Congress to address the issue in the context of other air quality management programs it has authorized. Sulfates, after all, are just another air pollutant.[4]

It is not clear, however, that using present CAA authorities to combat acid precipitation would neutralize the controversies in the Congressional arena. The combination of scientific uncertainty and differential regional impacts of any SO_2 control program have thus far forestalled action beyond the authorization of research (Acid Precipitation Act of 1980). Any commitment to abate acid precipitation would seem to require amending the CAA in some way-- whether to force action on a sulfate standard or to require emissions rollbacks from existing sources. Thus, even if forceful interpretation of existing CAA authorities and strong administrative action might effectively abate acid precipitation, current political realities make such actions unlikely.

The view of present policymakers was recently noted in a report which cited Charles Elkins, former head of EPA's Acid Rain Task Force and now director of

EPA air program development: "Congress never intended the EPA to solve the acid rain problem through the Air Act. Instead of trying to use the statute that way, Elkins said that the question of whether controls should be implemented should be settled 'straight on' in Congress."[5] And in a press conference announcing that EPA was denying the Pennsylvania, New York, and Maine petition for relief from acid precipitation under section 126, EPA Assistant Administrator for Air and Radiation Joseph A. Cannon was reported to have voiced the view that "the agency believes that Section 126 cannot be used to address problems posed by acid rain.... EPA will only have authority to control acid rain if Congress passes legislation giving the agency that authority, he said."[6]

New Source Performance Standards

The CAA does contain mechanisms that do have a bearing on the control of gross environmental loadings of air pollutants. Of these, the two most important for the acid precipitation issue are CAA requirements imposed on new sources and CAA requirements to prevent the significant deterioration of air cleaner than required by NAAQS. The "Prevention of Significant Deterioration" (PSD) program has the effect of controlling the growth of new sources but has only limited impact on reducing existing sources. The New Source Performance Standards (NSPS) program, while controlling new sources, can affect the emissions base as older facilities are retired. Some have argued that these requirements alone would abate SO_2 sufficiently over time to mitigate acid precipitation problems.

NSPS are emission limitations imposed on designated categories of new sources; and these limits apply even if the ambient air is and remains cleaner than would be required by NAAQS. After 1971, many high-sulfur coal-burning facilities met this requirement (as well as SIP requirements necessitated by NAAQS violations) by choosing low-sulfur rather than high-sulfur fuels even when the former were more expensive.[7] In the late 1960s, residual fuel oil was replacing coal for electricity generation on the East Coast because it was lower in cost and had the additional benefit of being lower polluting. With the CAA, pollution control requirements reinforced the trend toward low-sulfur fuels, even though the costs of these fuels were rising.

However, events soon made choosing low-sulfur fuels primarily for environmental reasons an unattractive national policy. Following the 1973 oil embargo, the Energy Supply and Environmental Coordination Act of 1974 prohibited facilities from switching from coal to oil or gas. Then, to limit shifts from abundant eastern high-sulfur coals to low-sulfur coals, the 1977 CAA Amendments required coal-burning utilities to meet a continuous, technology-based emission limitation. The resulting NSPS regulation for coal-fired utilities requires that emissions from high-sulfur coals be reduced 90 percent and emissions from low-sulfur coals be reduced 70 percent.[8] This requirement substantially reduces the total environmental loadings of SO_2 from the Nation's single largest category of SO_2 emissions--the coal-fired electric generating facilities--to the extent that facilities meeting NSPS replace existing ones. At present, only some 10 percent of utility emissions are from NSPS facilities. Only new sources (for present standards, those built since 1978) and "reconstructed" ones must meet the standard. (Reconstruction triggers NSPS when the fixed capital costs of the new components exceed 50 percent of the fixed capital costs of a comparable replacement [40 CFR 60.15])

In recent years it has been suggested that the NSPS provisions of the CAA were conceived as a way of decreasing emissions over time. In this view, NSPS solve pollution problems as new facilities replace existing, dirtier ones, just as cleaner, catalytic-converter equipped automobiles would replace older, more polluting models. Most electric power generating and coal industry interests have cited NSPS as a key reason for their opposing regulations to address acid precipitation problems. The Alliance for Balanced Environmental Solutions has written that, when one considers the evolving impact of NSPS, "it is apparent that the goals for sulfur dioxide in the Clean Air Act will be realized in the not so distant future."[9] Carl Bagge, President, National Coal Association, testified before Congress:

> The Clean Air Act is working. Under the Congressional prescribed New Source Performance Standards, new more efficient coal-fired plants are replacing older plants as they are operated at reduced capacity or retired. Cleaner coal and lower-sulfur coals are also being used. As a result, sulfur dioxide emissions have been reduced dramatically and this reduction will continue without any new legislation.[10]

In addition, graphic depictions of declining emissions due to NSPS have been prepared showing how NSPS alone will reduce utility SO_2 emissions to a level equal to those achieved from retrofitting existing sources by about 2015.[11]

Meanwhile, those who urge initiating control actions have suggested that, while NSPS were designed to reduce emissions, they are not working that way. In testimony before the Senate Environment and Public Works Committee, Dr. Ellis Cowling of North Carolina State University said:

> The assumption has been that the older plants would be gradually phased out as these facilities became more and more obsolescent. This phasing out was expected to result in a progressive decrease in emissions. New facilities (with [NSPS] controls in place) would gradually replace old facilities (without controls). These provisions of the Act in effect made the assumption that the amount of reduction necessary to protect public health, ecosystems, surface waters, and materials could be achieved primarily through management of emissions from new sources rather than some combination of both new and older sources.[12]

The legislative history of NSPS, however, does not mention this assumption.[13] In fact, through most of the 1970s, it appeared that even with NSPS and increased reliance on nuclear power, "the best that can be hoped is a total sulfur oxides emission rate from all utilities somewhere near the present [1970] level."[14] For it was assumed that electrical demand and generation would increase several times by the mid-twenty-first century.

Today, however, the situation is vastly different. First, projections of the future growth of electricity generation have been greatly reduced, and the use of coal is expected to accelerate as the nuclear industry continues to contract. Second, present emissions of SO_2 are viewed by many as causing acid precipitation at unacceptably dangerous levels. As a consequence, attention has shifted from the question of how to moderate future SO_2 emissions levels to the question of how to reduce existing emissions.

The convergence of these two developments creates a new situation spotlighting the potential role and effectiveness of NSPS provisions in reducing acid precipitation causing pollutants. When it was assumed that electricity generation would quadruple between 1970 and 2000, the contributions of existing sources to future emissions problems were seen as minor. From today's perspective, it is the old, dirty sources that constitute the crux of the problem; and, if acid precipitation is viewed as at least a potential threat now, it will very likely be seen as even more of a problem is any increase in emissions occurs in the future. So in the present context the debate has focused on the issues of how to abate emissions from existing facilities and whether the natural turnover of facilities will proceed fast enough to make a massive retrofit control program unnecessary. As previously noted, one set of projections arguably supports the view that emissions will decline continuously and substantially, while other projections suggest that NSPS requirements will not abate emissions as much and as quickly as necessary.

IMPLICATIONS

Legislatively, this review of CAA provisions has consequences for the positions of both proponents and critics of regulating acid precipitation pollutants. First, it appears that the CAA's provisions for controlling the ambient air quality will not effectively engage the acid precipitation problem. While in theory they might work, in practice the task of implementing them creatively and in a manner conducive to acid rain mitigation is overwhelming. Moreover, under prevailing political circumstances, it is impossible politically since the Reagan administration and its environmental protection arm are taking the position that the existing CAA will not be so implemented without explicit congressional directive. For proponents of immediate regulatory control on SO_2 to abate acid precipitation, this requirement for congressional action defines the arena of policy debate. It further means that their goal cannot be achieved without affirmative action, while the goal of the opponents of an immediate program is served by retaining the status quo.

Second, it appears that the CAA's NSPS provisions for controlling emissions from new sources may play an unexpected role. Designed to control the rate and distribution of new emissions, they could become the mechanism for reducing emissions. Any control program will take several years to implement. It can be argued, however, that replacement or reconstruction of existing, dirtier facilities with NSPS-meeting facilities could occur fast enough to make a massive retrofit program unnecessary. From the point of view of opponents of immediate action, relying on NSPS to bring down emissions has the advantage of not requiring a special, expensive retrofit program. Instead, the electric generating industry could evolve in a more normal way. For proponents of immediate action, reliance on NSPS may have the disadvantage of delaying the reduction in emissions. On the other hand, it does not require legislative action, which would be an advantage if NSPS will reduce emissions at a reasonable rate.

Obviously, the key questions are how much and how fast would NSPS requirements abate the emissions acting as acid precipitation precursors. If the amount and timing are such that retrofitting brings few gains, this would add weight to the arguments of the opponents of immediate regulatory action. But, if the reductions would be small and would only be achieved in the distant future, this would add weight to the arguments of proponents of action.

How fast will old facilities be replaced? The next section of this paper explores the forces affecting the rate at which NSPS might effectively reduce emissions in order to see what possible levels of emissions are likely if no legislative action is taken soon to impose a new regulatory scheme.

UTILITIES, GROWTH, AND FUTURE EMISSIONS

If the effectiveness of the CAA and NSPS in controlling future emissions is to be assessed accurately, one must estimate those future trends. Unfortunately, forecasting the future is akin to driving a car by looking through the rear view mirror: one is tempted to assume that what occurred in the past will occur in the future. To the extent one deviates from this historical "data," one runs the risk of being dismissed as a "maverick," "dilettante," or ill-informed tyro regardless of how well-reasoned the scenario might be. For the purposes here, we will provide an overview of the variables which influence future SO_2 emissions, a sketch of various perspectives on the future, an analysis of several alternative futures, and an assessment of their differing implications for the acid rain debate.

Variables Influencing Future SO_2 Emissions

At least five variables influence SO_2 emissions: electricity generation growth, availability of non-fossil-fuel-fired generation, sulfur content of fuel burned, lifespan of existing, uncontrolled generating facilities, and regulations.

Regulation. Over the last decade, regulation of SO_2 emissions has had a significant impact on the growth of those emissions. The decline in emissions in the mid-1970s is largely attributed to implementation of SIPs. While most of the SO_2 reductions during the 1980s have been due to less electricity generation, the impact of the 1971 and 1979 NSPS will continue to

TABLE 5.1
Impact of regulation on SO_2 emissions

% Sulfur in Coal*	Uncontrolled Emissions	1971 NSPS	1979 NSPS	95% Reduction
		(lb./mmBtus)		
0.5	1.2	1.2	.4	.06
1.0	1.8	1.2	.2	.09
2.0	3.6	1.2	.4	.18
3.0	5.5	1.2	.6	.27

*For the .5 percent coal, 8,500 Btus/pound; for all others, 11,000 Btus/pound.

FIGURE 5.1
Impact of electricity demand growth on SO_2 emissions

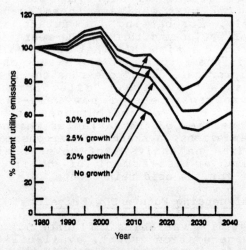

*Assumes 50-year retirement age

Source: Office of Technology Assessment, <u>Acid Rain and Transported Air Pollutants</u>, OTA-O-204, June 1984, p. 62.

affect the growth of emissions in the future.

The potential impact of regulation on emissions is indicated in Table 5.1. Obviously, the initial move in 1971 to implement a 1.2 lb. SO_2 mmBtu limit on emissions from new coal-fired powerplants resulted in the greatest reduction in allowable emissions from future power plants, since they were previuosly uncontrolled. This significant reduction in emission rates for future plants was increased an additional 50 percent or more by the revised 1979 NSPS. Nevertheless, since NSPS requirements apply only to new plants, emissions are likely to stay relatively high as long as old plants predominate. (The implications of their retirement rate is discussed later.) Indeed, in 1980, older, non-NSPS facilities were still emitting SO_2 at a rate exceeding 10 lbs. of SO_2 mmBtus. Since States have found reductions necessary to meet NAAQS, any controls imposed on their older plants would be through SIPs.

As new technologies capable of 95% to 99% removal of SO_2 are developed, Congress or the EPA may decide to strengthen the SO_2 NSPS even more. Indeed, the Congress may decide to "force" the accelerated development of this technology by enacting such a requirement a few years earlier than the planned commercialization of these technologies. The last column of Table 1 indicates the emission rates from powerplants meeting a 95 percent removal requirement. As indicated, such a standard would result in at least a 50 percent reduction in allowable emissions from the stringent 1979 standards. The impact of such a standard from previous uncontrolled plants is more dramatic with emissions only five percent of their uncontrolled levels.

Electricity Generation Growth: Electricity generation growth affects SO_2 emissions in two ways: (1) to the extent that the new generation source is fossil-fuel-fired, it raises emissions directly; and (2) to the extent that it is a response of growing economy, other SO_2 emitting sectors will elevate emissions. The impact of this factor was inversely illustrated in the early 1980s as SO_2 emissions dropped significantly in response to reduced demand for electricity during the recession.

Figure 5.1 illustrates the potential impact electricity generation growth can have on future emissions. As indicated, assuming just a 2 percent/year growth in electricity demand results in major reductions in emissions during the first quarter of the 21st century under the assumptions of the analysis. However, if electricity growth were 3 percent/year, a major reduction in emissions would not be achieved. If the increase were even greater, then

FIGURE 5.2
Domestic nuclear capacity, 1982-2020

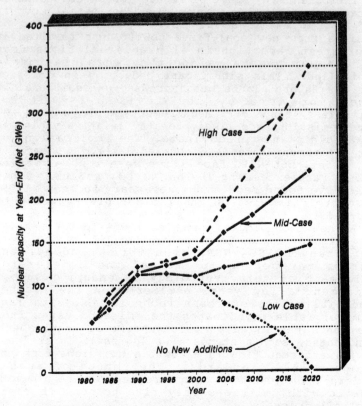

Source: Energy Information Administration

TABLE 5.2
Impact of Alternative Generating Sources on SO2 Emissions (150 Gigawatt Scenario)

Emission Rate of Fossil-fuel fired Generation Displaced (lb. per mmBtus)	SO2 Removed (millions of tons)
1.2	4.7
2.84	11.2

no reduction might occur at all.
Availability of Non-Fossil-Fuel-Fired Electricity.
One way to cut emissions is to reduce reliance on fossil-fuel-fired electricity (coal and oil). During the late 1960s and early 1970s, shifting to nuclear power was seen as a key to holding down projected increases in SO_2 emissions. Now, in the 1980s, others have suggested that energy technologies such as solar power might take over the role of reducing the burden of electricity generation currently shouldered by coal.

The potential and uncertainties regarding the future of non-fossil-fuel-fired generation are illustrated by current projections for nuclear power. As shown in Figure 5.2, the range of nuclear capacity in 2020, as projected by the Department of Energy is 350 gigawatts.[15] Projections for other non-fossil-fuel-fired generation sources (with the exception of hydropower) are just as uncertain. This uncertainty translates into millions of tons of SO_2 which may or may not be produced because of the availability of non-fossil-fuel alternatives.

An example of the potential impact of these alternatives is shown in Table 5.2. As indicated, a 150 Gs change from fossil-fuel generation to an alternative source (or a shift from the no-new-additions case to low case in Figure 5.2) would result in a reduction of between 4.7 and 11.2 million tons, depending upon the emission rates of the capacity displaced. Such an ability could be very important in the future as the demand for electricity increases.

Retirement of Existing Facilities. Until very recently, the retirement of existing facilities was not considered an important variable by most analysts. Powerplants were assumed to last their designed lifespan of about 30 years and then be replaced by newer facilities. Indeed, it has been assumed that the CAA's requirement that new facilities meet NSPS would over time result in the air becoming cleaner as these less-polluting plants replaced older, dirtier plants. Over the last five years, however, it has become apparent that the actual lifespan of powerplants is not set, but is instead quite elastic. With new powerplants costing over $1000 a kilowatt to construct, utilities have powerful incentives to avoid construction and to rehabitate older facilities. This incentive is partially reinforced by environmental regulations which permit facilities to be rehabitated up to 50 percent of their assessed value without being required to meet NSPS (i.e., scrubbers). With the cost of such rehabitation estimated at about $500 a kilowatt (although that number can vary substantially), operating existing facilities for upwards of 60 years seems a likely trend in the future.

FIGURE 5.3
Estimated emissions as a function of retirement age of powerplants

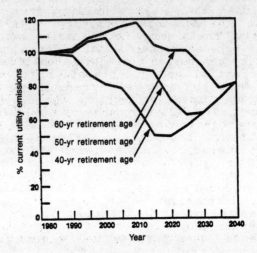

*Assumes 2.5% a year growth in electricity demand.

Source: Office of Technology Assessment, <u>Acid Rain and Transported Air Pollutants</u>, June 1984.

FIGURE 5.4
Coal Sulfur delivery trends for electric utilities in
selected census regions and in the United States,
1976-1982

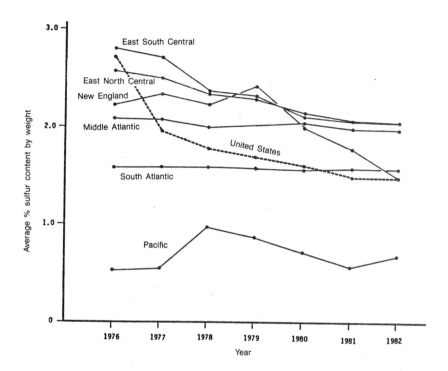

Source: Applied Management Sciences, Inc. Coal and
Oil Deliveries to Electric Utilities in Each State,
1976-1982. Prepared for U.S. Department of Energy,
1983, draft report.

Figure 5.3 illustrates the impact the extension of existing plant life can have on future SO_2 emissions. As shown, if existing plants are assumed to have a "standard" life of 40 years, significant reductions in SO_2 emission would begin occurring around the year 1990 under the assumptions of the analysis. However, if the life of these plants is extended to 60 years, reductions do not occur until the second quarter of the 21st century. Indeed, the reduction achieved is rather modest and short-lived.

Sulfur Content of Fuel. Perhaps the most straightforward variable involved in SO_2 emissions is the fuel's sulfur content. Indeed, switching facilities to lower sulfur coal was a primary means which utilities employed in the mid-1970s to achieve SIP requirements. This is illustrated in Figure 5.4.

With utilities extending the lifespan of plants which do not have other technological SO_2 controls, sulfur content could remain an important variable in determining absolute totals of SO_2 emissions in the future. Indeed, as shown in Table 5.1 earlier, even with technological controls, the sulfur content of the coal still has some relative influence over total emissions--although its absolute impact is much less.

Perspectives on the Future

As suggested by the above discussion, one can develop any future one wants by "adjusting" the appropriate variables. For example, one possible scenario, shown in Figure 5.5, has been used by some groups to argue that the basic framework of the CAA is adequate to handle acid rain precursors and no additional controls are necessary.[16] This conclusion, however, is the result of critical assumptions with respect to the five variables discussed above (as are all projections of future emissions trends). Specifically, the projection makes two assumptions: (1) a lifespan of about 40 years for existing uncontrolled facilities, ensuring that the basic logic of the CAA of new cleaner plants replacing older, uncontrolled plants occurs; and (2) a viable nuclear option, ensuring that 25 to 30 percent of total generation will be non-fossil-fuel-fired and, therefore, non-SO_2 emitting. For the new technology scenario, it is assumed that the plants employing such technology will choose to achieve a 95 percent removal of SO_2.

Reasonable people may differ on the reasonableness of the above assumptions. However, for a Congress concerned about the risks of prolonged emissions of current levels of SO_2, it is important to note that there is a significant probability that economic and

125

FIGURE 5.5
One alternative view of future utility emission trends

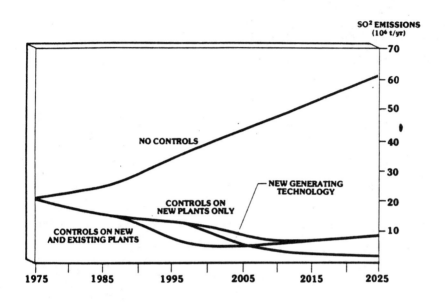

Source: EPRI Journal, November 1983.

other forces will prevent the above scenario from occurring on its own. Current trends in powerplant construction economics indicate that existing plants are and will continue to have their useful lives extended beyond 40 years, and nuclear power faces an uncertain future. Hence, while the above scenario could happen (particularly with government intervention), it is by no means an inevitable outcome of present conditions concerning economic-energy growth, technological developments, regulatory requirements, and other variables.

In order to examine future trends in SO_2 emisisons based on current economic trends in plant construction, we have developed alternative scenarios which indicate both the range of futures possible and some alternative policy options.

Methodology

The object of this analysis is not to predict the future, but to illustrate and compare the impact of different options given a base set of assumptions. It should also be pointed out that this analysis discusses SO_2 emissions from utility plants only, and thus excludes emissions from industrial sources.

Four scenarios were developed for this analysis: (1) a 60-Year, Current NSPS--this assumes a 60-year lifespan for existing facilities and current NSPS for future plants; (2) a 45-Year, Current NSPS--this assumes a 45-year lifespan for existing facilities and current NSPS for future plants; (3) a 60-Year, Revised NSPS--this assumes a 60-year lifespan for existing facilities and a revised 95 percent removal NSPS for future plants; and, (4) a 45-Year, Revised NSPS--this assumes a 45-year lifespan for existing facilities and a revised 95 percent removal NSPS for future plants.

The object of this exercise is to vary those parameters which the Federal Government already has significant authority to control. Obviously, the government has direct control over regulation of future plant emissions. Also, through environmental regulations, such as the 50 percent modification rule, the effective lifespan of existing facilities can be greatly influenced. The other variables over which government has less control are assumed to follow currently projected paths to 2030 with nuclear power construction assumed to stop in 1995 and with existing nuclear plants retired after 30 years of life. Table 5.3 indicates the supply mix assumed in the analysis.

The 45-60 year lifespan reflects the difference between the traditional lifespan of a powerplant (and part of the basis of the CAA) and current expectations of a practical extension of current powerplant life

TABLE 5.3
Forecast of available electric generating capacity, 1983-2030, probable case 1/

Item	1983 Actual	1995	2000	2005	2010	2015	2020	2025	2030
Net Energy (Billion Kwh) 2/	2310	3346	3888	4517	5245	6091	7069	8206	9523
Peak (Million Kw) 3/	448	597	694	806	936	1087	1261	1464	1699
Resources Needed (Million Kw) 4/	596	716	833	967	1123	1304	1513	1757	2039
Type of Capacity (Million Kw)									
Coal	256	317	454	626	787	96	1159	1396	1657
Gas	38	30	15	—	—	—	—	—	—
Hydro	69	73	74	76	78	79	81	82	84
Nuclear	60	101	97	78	59	51	41	21	—
Oil	42	31	15	—	—	—	—	—	—
Nonconventional	1 5/	3 5/	5	8	13	20	33	52	85
Other	130 6/	144 6/	150	156	163	169	176	183	190
Total	596	699	810	944	1100	1281	1490	1734	2016
Imports (Million Kw) 7/	10	23	23	23	23	23	23	23	23

1. Assumes 3 percent real annual GNP growth, 0.8 percent real increase in price each year, and a 64 percent LF.

2. EG = -19 + (1.56 GNP) - (1629 Average Revenue).

3. Peak = EG/(64% x 8760 hrs.)

4. Assumes 20 percent reserve margin.

5. Geothermal only.

6. Includes some nonconventional sources.

7. Converted to equivalent capacity at average capacity factor for Canadian Hydro (60.9%).

TABLE 5.4
Assumptions for four scenarios

Scenario	Regulation	Existing Plant Lifespan	Variable			
			Electricity Growth	Available Non-Fossil Generation	Sulfur Content of Fuel
60-Year, Current NSPS	Assumes continuation of current NSPS	60 Years	3.1 Percent annual increase based on a 3 percent growth in GNP. (Same for all scenarios)	No new nuclear plants constructed after 1995. Constructed nuclear and oil and gas plants assumed to have 30-year lifespans. Renewables rise to 5 percent of peak in 2030. Hydro and other rise at their 1983-95 rate as projected by NERC. Imports assumed level at their 1995 rate as projected by NERC. (Same for all scenarios)	1.5 percent, 10,500 Btu/lb. for all coal except 1971 NSPS coal is assumed to be .5 percent sulfur, 8,000 btus/lb. Oil is assumed to be 1 percent sulfur 60 percent capacity factor as assumed for coal plants, 30 percent for oil. Heat Rates: Un controlled; 10200 Btus; 1971 NSPS, 10300 Btus; FGD, 105 Btus; Advanced Coal, 9300 Btus (Same for all Scenarios)
45 Year, Current NSPS	Same	45 Years			
60-Year, Revised NSPS	After 1995, new plants assumed to meet a 95 percent removal requirement.	60 Years			
45-Year Revised NSPS	Same	45 Years			

(indicated by the current trend in powerplant construction). The two regulation scenarios reflect an extension of the current NSPS and the establishment of a more stringent NSPS (95 percent removal) as new, more efficient technology becomes commercially available. Also, the revised NSPS scenario could reflect a scenario with a more optimistic future for nuclear power, thus placing less generating burden on coal. Table 5.4 details the assumptions of these scenarios in terms of the five variables identified earlier.

The current NERC projections in 1995, as modified by Kaufman, et. al.,[17] is the base year of the analysis.[18]

Results

Figure 5.6 presents the results of the analysis. Not surprisingly, a wide range of futures is indicated. The 60-year, current NSPS scenario, reflecting current trends in plant construction, shows a steady increase in emissions through the time analyzed. This results because, by the time existing plants are retired, the need for additional capacity (due partially to the phase-out of nuclear) leads to increasing emissions. In the 45-year, current NSPS scenario, the retirement of existing uncontrolled facilities results in a slight downward trend; but again the increasing need for electricity drives emissions up toward the end of the time period.

The two revised NSPS scenarios result in fundamentally different trends. In the 60-year, revised NSPS scenario, emissions peak around 2010 and begin a significant decline as old plants are replaced by advanced plants. This circumstance occurs despite the phase-out of nuclear power. In the 45-year, revised NSPS scenario, this downward trend begins immediately in 1995 and continues until 2025 when the phase-out of nuclear is complete and the increase in electricity demand causes emission levels to stabilize.

The 60-year scenarios probably reflect current trends in plant construction. If this hypothetical future proved correct, then significant reductions in current levels of utility emissions would be unlikely until well into the 21st century, if then. However, the 45-year, revised NSPS scenario illustrates that this does not have to be the case. If the government chooses to limit modifications to existing plants or to lower the threshold for meeting NSPS for such modifications and promulgate a stringent NSPS or encourage use of non-fossil-fuel generation, a downward slope is possible. However, positive action by government is probably necessary to obtain it.

130

FIGURE 5.6

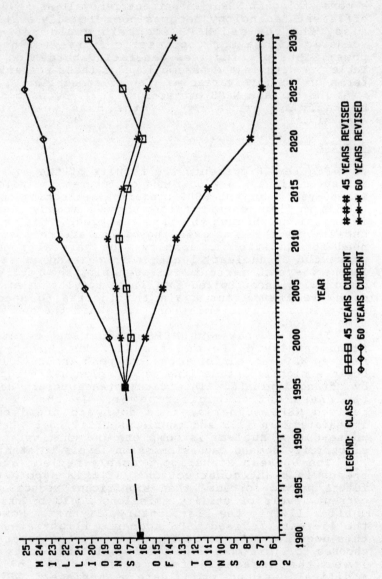

Implications for Acid Rain

The implications of the previous discussion for acid rain depends on one's view of the urgency of the problem. For those who believe that reductions must begin immediately with significant reductions (i.e., 8-12 million tons) by 1990-2000, the analysis has few implications for their basic strategy in the short term: an immediate retrofit program is necessary. On other hand, for those proponents also concerned about the long-term emissions of SO_2, and for those who feel acid rain control is not as urgent, the analysis has several implications. Those implications are illustrated in Figure 5.7 and may be seen by comparing the two revised NSPS scenarios with two retrofit scenarios: (1) a 60-Year, Current NSPS, 10-million-ton Retrofit Program; and (2) a 60-Year, Revised NSPS, 10-million-ton Retrofit Program. As indicated, without more stringent controls on future plants or increased use of non-fossil-fuel-field generation, a 10-million-ton reduction in SO_2 could be temporary. With few uncontrolled plants retiring to offset future growth or a more stringent NSPS, emissions would tend to climb in proportion to new fossil-fuel-fired generation. Hence, while a retrofit program controls existing emissions effectively, it does not guarantee emissions would remain low over the long run--unless, as has been proposed in some bills, future emissions were capped.

This suggests several policy alternatives. One alternative is illustrated by the second retrofit program combining a retrofit program on existing facilities with a more stringent NSPS on future facilities. As shown, this combination would tend to keep emissions down throughout the time period--despite a 60-year lifespan for facilities, future growth in generation, and the phase-out of nuclear power.

A second alternative is suggested by the 45-Year, Revised NSPS scenario. This scenario would require government action to shorten the lifespan of uncontrolled facilities and a revised NSPS, but would avoid the substantial expense of a retrofit program. This saving would come at the expense of an additional 15 to 25 years of significant levels of SO_2 emissions. Whether this represents an unacceptable risk to the environment depends, as stated earlier, on one's perception of the problem's urgency.

CONCLUSION

The politics of acid rain are immensely complicated. Technically, the problem's seriousness, its urgency, and its susceptibility to present control

FIGURE 5.7

SIMULATIONS OF TWO RETROFIT PROGRAMS WITH TWO REVISED NSPS PROGRAMS

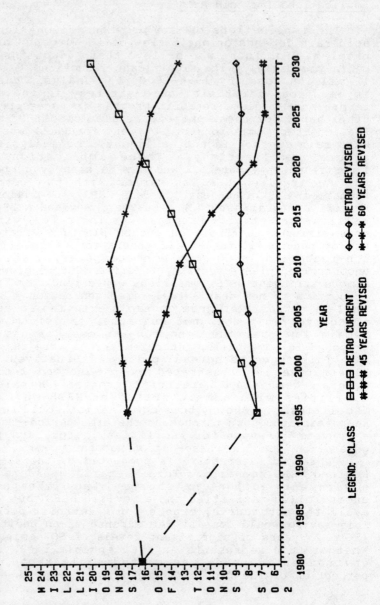

capabilities are debatable, and intensely so because the costs implied by most control programs are extremely high. Further, the costs and benefits are not distributed uniformly, but rather have the effect of pitting different regions of the country against one another. Powerful economic sectors of the nation have vested interests. Some programs imply substantial social changes, especially for coal-producing regions. In addition, the issue is inextricably bound up with other Clean Air Act issues, which are themselves highly controversial.

As discussed in the first part of this chapter, several provisions of the Clean Air Act provide a basis for tackling the problem of acid precipitation. EPA has rejected such aggressive interpretations of the Act, and its air quality program administrators have stated that action now--that is, pending further results of research--would require explicit Congressional direction.

For a variety of reasons, such Congressional action seems doubtful in the near future. First, the uncertainties about the problem, how it might be controlled, and ultimate control costs provide considerable leverage to opponents of a legislated regulatory program. Second, the differential regional impacts of various bills pose a particularly difficult barrier to finding a compromise position. Third, the opposition of the Administration to a regulatory program means that, not only is there no supraregional leadership to resolve regional differences, but also that even if Congress found a compromise program it might be vetoed.

A final force obstructing prompt Cngressional enactment of an acid rain program is the claim that a convergence of trends in electrical energy growth and future emissions will combine with existing CAA requirements--the New Source Performance Standards--to cause an ineluctable downward trend in SO_2 emissions. Indeed, since 1970, when the foundations of the present CAA were enacted, SO_2 emissions have declined by some 25 percent. Opponents of an acid rain regulatory program contend that this trend will inevitably continue, and that any immediate retrofit program would be extremely expensive and would only serve to hasten the reductions slightly.

In this view, the replacement of old facilities with NSPS-meeting new plants will over time reduce gross emissions. This view is based on the belief that the absolute growth in electrical generation will decline to such an extent that the replacement of older, more polluting facilities by new, cleaner ones would more than offset required expansion of capacity. The argument sees an analogy in the control of auto pollutants. That is, existing cars were not retrofitted; instead, as new, catalytic-converter equipped autos gradually replaced older ones, emissions declined. For anyone seeking to avoid the costs of retrofitting existing sources to reduce SO_2 emissions in order to mitigate acid precipitation damages, the analogy is compelling if true, seductive even if not.

The analysis in the second part of this chapter suggests that the analogy is most likely false. It suggests, instead, that SO_2 emissions will most likely not continue to decline under circumstances influenced by the Clean Air Act regulations and by probable energy trends. The primary reason that NSPS will not have the imputed effect of causing future declines in SO_2 emissions of coal-fired utilities is that older, dirtier plants will most likely not be retired or reconstructed at historic rates, but instead will be refurbished and kept in service for years beyond their original design life. Such plants will not have to meet the more stringent NSPS requirements; indeed, the costs of NSPS only increase the economic attractiveness of not replacing older facilities, given existing technology. It is this ability to extend the life of fossil-fuel-fired-steam-generating facilities that negate the analogy to automobiles: the turnover rate of automobiles is much faster and much less extendable than the turnover rate of utilities boilers.

This analysis has two major consequences for the acid rain debate. First, if the analysis is correct, then the claim that acid precipitation controls are unnecessary because emissions of SO_2 will be declining anyway is not sound, and any decision on whether or not such controls are warranted today should be based on other considerations. Second, if emissions in the future do begin to rise, the present debate over a retrofitting program may be overtaken in a few years by a debate over how to prevent emissions growth. Such a shift is particularly likely if, as hypothesized, Congress does not enact an acid precipitation control program in the next few years.

By isolating the key role the turnover rates of existing sources play in determining future emissions trends, this analysis suggests that one focus of the acid precipitation debate should be on how long a facility can remain in service without having to meet NSPS. Congress could legislate on this point, although for reasons discussed earlier CAA amendments may not be likely for several years. Alternatively, Administrative action could be taken to change the definition of "reconstruction," so it would be more difficult to refurbish a facility without having to meet NSPS. EPA's present disinclination to address acid precipitation through present authorities of the

CAA might well be overwhelmed if SO_2 emissions actually begin rising significantly. Focusing on the likely future role of old, non-NSPS facilities might lead to other alternatives. For example, it might be effective and appropriate to separate emissions requirements imposed on baseload and on peaking-load facilities--requiring the former to meet more stringent standards, whether new or old. Finally, the anlaysis indicates that increasing the stringency of the NSPS may be necessary to keep SO_2 emissions trends down in the future, particularly if the nuclear option remains moribund.

NOTES

1. Margret R. Gallogly, "Acid Precipitation: Can the Clean Air Act Handle It?" **Environmental Affairs**, 9 (1981), p. 687; Eugene M. Trisko, "The (Non)Integration of Law and Policy: Acid Rain Control Meets the Clean Air Act," **Environmental Forum**, Vol. 2 (October 1983), pp. 12-13.
2. "Tightening Compliance Policy Could Reduce Sulfur Dioxide Emissions But Cost to Midwest Would Be High, State, EPA, Industry Officials Say," **Environment Reporter**, March 16, 1984, pp. 2059-2064.
3. Eugene M. Trisko, "Speaking Out--Acid Rain Remedies and Clean Air Act Revisions," **Inside E.P.A.** March 23, 1984, p. 16. Also, Trisko, "Controlling Acid Rain with the Clean Air Act," **Environmental Forum**, in press.
4. **Ibid.**
5. "Tightening Compliance Policy Counld Reduce Sulfur Dioxide Emissions But Cost to Midwest Would Be High, State, EPA, Industry Officials Say," **Environment Reporter** (March 16, 1984), p. 2064.
6. "EPA Denies Northeast States' Petitions Asking for Action on Emissions in Midwest," **Environment Reporter**, December 7, 1984, p. 1327.
7. After August 17, 1971, the coal-fired electric utility NSPS emission requirement was 1.2 pounds SO_2 per million Btu (40 CFR Subpart D).
8. This requirement went into effect September 18, 1978 (40 CFR Subpart Da). Although the "percentage reduction" requirement of the CAA applies "fossil-fuel-fired stationary sources" generally, the regulations issued apply only to large electric utility steam generating units and do not apply to industrial boilers.
9. Alliance for Balanced Environmental Solutions, "Common Sense about Clean Air: A Thoughtful Look at Acid Rain" (Washington, D.C.: Alliance for Balanced Environmental Solutions, 1984).

10. Testimony in U.S. Congress, Senate: Committee on Energy and Natural Resources. Implementation of the Acid Precipitation Act of 1980, hearing, 98th Cong., 2d sess. [S. Hrg. 98-963] (Washington: U.S. Govt. Print. Off., 1984), p. 724.

11. Ralph Whitaker, "Developing the Options for Emissions Control," **EPRI Journal**, November 1983, p. 50.

12. Ellis B. Cowling, "Four Questionable Assumptions Underlying the Clean Air Act of 1970 and Its Amendments of 1977," in U.S. Senate, Committee on Environment and Public Works, **Acid Rain, 1984**, hearings, February 2, 7, 9, and 10, 1984. [S. Hrg. 98-714] (Washington: U.S. Govt. Print. Off., 1984), p. 201.

13. The legislative history justifies NSPS on the grounds of (1) precluding States from competing for new industry by having laxer standards; (2) enhancing the potential for long-term growth; (3) saving costs by avoiding any future need to retrofit controls; (4) ensuring competitiveness between low- and high-sulfur coals; and (5) creating incentives for new control technologies. (Senator Edmund Muskie, Senate Consideration of the Report of the Conference Committee, August 4, 1977, in U.S. Senate, Committee on Environment and Public Works, **A Legislative History of the Clean Air Act Amendments of 1977**, [95th Cong., 2d sess.; Serial No. 95-16] (Washington: U.S. Govt. Print. Off., 1979), Vol. 3, p. 353.

14. Ad Hoc Panel on Control of Sulfur Dioxide from Stationary Combustion Sources, National Research Council, **Abatement of Sulfur Oxide Emissions from Stationary Combustion Sources** (Washington: National Academy of Engineering, 1970), p. 14-15.

15. Energy Information Administration, "Commercial Nuclear Power: Prospects for the United States and the World," (Washington, D.C.: U.S. Govt. Print. Off., n.d.).

16. Alliance for Balanced Environmental Solutions (an industrial group), "Common Sense About Clean Air: A Thoughtful Look at Acid Rain," op.cit.

17. Alvin Kaufman; Donald Dulchinos; and Larry Parker. "Gold at the End of the Rainbow? A Perspective on the Future of the Electric Utility Industry," Congressional Research Service, December 1984.

18. North American Electric Reliability Council, "Electric Power Supply & Demand, 1984-1993" (Washington, D.C.: N.A.E.R.C., 1984).

6
Predicting Deposition Reductions Using Long-Range Transport Models: Some Policy Implications

Glenn P. Gibian

INTRODUCTION

Several pieces of legislation which would require significant reductions in sulfur dioxide emissions are under consideration in the Congress. These bills are viewed by their sponsors as necessary to control acid rain in certain sensitive areas, such as the Adirondacks, New England and southern Ontario. Evaluation of expected benefits from the legislative proposals however has been limited. Normally, in constructing environmental control legislation, one first determines the environmental changes which are desired and then determines what controls are necessary to achieve these changes (U.S. EPA Acid Deposition Task Force, 1983: 8). Thus, the expected benefits are known. The approach taken by Congress and others has been to set out emission control strategies and then hope that these will result in environmental benefits. There has been little scientific evaluation of what benefits would result from the strategies and what work has been done has received little attention by the public or policymakers.

This chapter attempts to evaluate and compare various emission control proposals. Further, it attempts to illustrate the significance of the two different approaches to developing environmental control legislation. Specifically, it demonstrates that strategies developed with specific environmental goals in mind will be more effective and efficient than the strategies presently being considered by Congress. Additionally, it seeks to evaluate the use of utility emissions for allocating emission reductions to be achieved by different states.

LONG RANGE TRANSPORT MODELS

Description

Long-range transport models attempt to simulate mathematically physical phenomena including the transport, diffusion, transformation and deposition of pollutants over long distances (U.S.-Canada Work Group 2, 1982: 46). The models may be used to predict the fate of pollutants emitted from various sources and, conversely, the origins of pollutants being deposited at receptors. Source-receptor relationships may then be quantified. This simulation is subject to considerable uncertainty as modelers themselves are the first to note. Techniques of simulation are varied since 1) the best technique is not clearly discernible; 2) the nature of all relevant physical and chemical processes is not well understood; 3) each modeler makes varying assumptions to simplify the complex processes; and 4) models are often developed independently (U.S.-Canada Work Group 2, 1982: 4).

The U.S.-Canada Work Group on Atmospheric Sciences and Analysis (Work Group 2) has published "transfer matrices" which represent source-receptor relationships for seven such models using 1978 meteorologic data. These are used here to estimate the effects of emission reduction strategies.

Use and Explanation of Transfer Matrices

While atmospheric models of long-range transport and transfer matrices are subject to considerable uncertainty and limitations, they may be the best tools available for predicting the effects of specific emission reduction strategies on specific receptor areas. Work Group 2 was charged with recommending tools for preliminary assessment activities including the estimation of emission reductions that would be needed in source areas in order to achieve proposed reductions in deposition rates necessary to protect sensitive areas. The Work Group stated that the principal tools available at present are air quality simulation models and transfer matrices. Further, transfer matrices are a convenient form for applying the results of the models to the study of emission reduction strategies (U.S.-Canada Work Group 2, 1982: 85).

An illustrative portion of a transfer matrix for wet sulfur deposition is included as Table 6.1. Each matrix contains transfer coefficients for 360 source-receptor combinations in units of kilograms (kg) of sulfur per hectare (ha) deposition per teragram (tg) of sulfur emitted. As an example, referring to Table 6.1,

TABLE 6.1

Example Portion of a Transfer Matrix

SOURCE	RECEPTOR			
	Algoma	Muskoka	Vt/N.H.	Adirondacks
N. Manitoba	0.45	0.25	0.20	0.29
S. Manitoba	0.77	0.77	0.00	0.00
N.W. Ontario	1.67	1.67	1.67	1.67
N.E. Ontario	2.99	0.91	1.43	1.04

Matrix elements are in units of kilograms of sulfur deposition per hectare at the receptor per teragram of sulfur emitted at the source

TABLE 6.2

Calculated Current Deposition Rates

Region	pH	H^+ (umole/liter)	Molar Ratio[c]	H^+ ($SO_4^=$)	H^+ (NO_3^-)
Adirondacks	4.2 [a]	63	1.0	42	21
Vermont/N.H.	4.26[a]	55	1.0	37	18
Muskoka	4.2 [b]	63	0.9	41	22
Algoma	4.35[b]	45	1.1	31	14

[a]U.S.-Canada Memorandum of Intent, Work Group 1 Impact Assessment, Final Report, Jan. 1983, p. 2-14

[b]The Case Against Acid Rain, A Report on Acidic Precipitation and Ontario Programs for Remedial Action, Ontario Ministry of the Environment, Oct. 1980 p.2

[c]National Research Council, National Academy of Sciences, Acid Deposition. Atmospheric Processes in North America, Washington, D.C. 1983, p. 130

if the source region Northern Manitoba emitted 1 teragram, it would contribute 0.45 kg of wet sulfur per hectare to the receptor Algoma, 0.25 to Muskoka, and so on. To determine the actual predicted contribution of each source area, it is necessary to multiply actual emissions by the coefficient. For example, source region Northern Manitoba emitted 0.4735 teragrams in 1978; thus it contributed (.4735) (.45) = 0.213 kg/hectare/yr to Algoma according to this model. The report contains separate matrices for wet sulfur deposition and dry sulfur deposition. The source regions are generally individual states, a group of states, provinces or parts of provinces; throughout this chapter, source regions will be referred to as "states".

For this chapter, each state's sulfur emissions were multiplied by the sum of its coefficients for wet and dry deposition for the receptor under consideration to determine each state's contribution to total sulfur deposition at the receptor. These contributions were totaled to represent total man-made deposition at the receptor under current emissions.

For several legislative proposals, each state's contribution (in kg./ha.) was reduced by a percentage equal to the percentage reduction in emissions required by the legislative proposal. The total of these reduced contributions then represents the reduced deposition at the receptor. This was done for each of the seven transfer matrices. For each model, the reduced total was then compared with the total under current emissions to represent percent reduction in deposition predicted by that model. Several hypothetical emission reduction strategies were evaluated similarly.

Finally, the reduction in wet sulfur deposition was calculated using the matrices for wet sulfur deposition and this was converted into reduced acidity of precipitation. For each site, the average 1979 or 1980 concentrations of $SO_4^=$ and NO_3^- were extrapolated from measured values reported in the literature as shown in Table 6.2. For each case of reduced wet sulfur deposition, the concentration of $SO_4^=$ in precipitation was reduced by the same percentage. NO_3^- concentrations were assumed to remain constant.

Comparison of Model Results Under Current Emissions

While there is significant variation between models, the Work Group stated that the extent of agreement among the model outputs is encouraging. Importantly, all models "predict generally similar relative impacts on the receptors in terms of ranked order of importance..." (U.S.-Canada Work Group 2,

TABLE 6.3

PERCENT REDUCTION IN SULFUR DIOXIDE (SO_2) EMISSIONS UNDER
VARIOUS CONTROL STRATEGIES
FOR THE UNITED STATES

State	Senate Bill	Sikorski/ Waxman	D'Amours Bill	Stafford	Uniform	Power Plant Elimination
Alabama	41.0	37.0	46.1	48.7	44.4	76.0
Arkansas	0.7	9.4	7.3	1.0	44.4	34.0
Connecticut	0.0	0.3	0.1	0.0	44.4	22.0
Delaware	18.7	21.4	20.8	26.4	44.4	58.0
D.C.	0.0	0.0	2.7	0.0	44.4	10.0
Florida	36.5	30.5	37.2	43.7	44.4	78.0
Georgia	54.0	54.0	57.8	64.5	44.4	87.0
Illinois	50.2	51.2	55.8	60.3	44.4	74.0
Indiana	57.3	58.4	63.1	68.7	44.4	74.0
Iowa	37.4	38.3	50.4	45.3	44.4	68.0
Kentucky	62.9	63.2	65.0	75.8	44.4	89.0
Louisiana	0.0	0.0	14.3	0.0	44.4	10.0
Maine	2.8	2.8	48.4	3.3	44.4	17.0
Maryland	33.3	31.9	34.7	40.0	44.4	68.0
Massachusetts	22.2	30.3	30.3	27.6	44.4	58.0
Michigan	25.6	25.7	34.8	30.9	44.4	61.0
Minnesota	22.8	24.3	32.2	28.6	44.4	61.0
Mississippi	37.3	25.2	31.4	42.8	44.4	67.0
Missouri	67.6	68.7	70.6	81.0	44.4	85.0
N. Hampshire	46.1	53.2	61.6	55.0	44.4	74.0
New Jersey	10.6	13.4	16.9	13.3	44.4	30.0
New York	21.0	23.5	28.2	25.6	44.4	45.0
N. Carolina	14.4	10.9	22.5	17.5	44.4	71.0
Ohio	57.8	60.1	70.1	69.4	44.4	77.0
Pennsylvania	40.5	41.8	46.1	48.8	44.4	67.0
Rhode Island	0.0	0.0	0.0	0.0	44.4	16.0
S. Carolina	30.7	28.9	37.2	38.0	44.4	70.0
Tennessee	62.5	61.9	66.0	75.1	44.4	86.0
Vermont	0.7	0.4	8.8	0.9	44.4	3.0
Virginia	8.0	7.2	20.7	11.0	44.4	50.0
W. Virginia	53.2	52.5	57.3	63.6	44.4	87.0
Wisconsin	49.8	53.7	13.2	60.1	44.4	72.0
(West)*						
W.N.E.	0.0	13.7	15.6	0.0	0.0	0.0
W.S.E.	0.0	7.1	16.4	0.0	0.0	0.0
W.N.W.	0.0	7.9	17.5	0.0	0.0	0.0
W.S.W.	0.0	0.0	17.7	0.0	0.0	0.0

*Western regions:
W.N.E. includes Nebraska, North Dakota, South Dakota, Montana, and Wyoming
W.S.E. includes Oklahoma, Kansas, Colorado, New Mexico, and Texas
W.N.W. includes Washington, Idaho, and Oregon
W.S.W. includes California, Nevada, Utah, and Arizona

TABLE 6.3 (continued)

PERCENT REDUCTION IN SULFUR DIOXIDE (SO_2)
EMISSIONS FOR CANADIAN SOURCES

Canadian Sources	Power Plant Elimination
Northern Manitoba	0.0
Southern Manitoba	10.3
Northwestern Ontario	39.2
Northeastern Ontario	0.0
Sudbury, Ontario	0.0
Southwestern Ontario	59.5
Southeastern Ontario	0.9
Montreal & St. Lawrence Valley	0.4
Norando & North Central Quebec	0.0
Gaspe Bay, Quebec	0.0
New Brunswick	56.8
Nova Scotia & Prince Edward Island	56.7
Newfoundland & Labrador	33.6
Saskatchewan & Alberta	12.1
British Columbia & Yukon	0.3

1982: 96). In other words, the models generally agree upon which source area will have the greatest impact on a given receptor, which source area will have second-greatest impact on the receptor, and so on. The models vary when quantifying how much greater will be the emissions from another source area. For example, all models agree that that emissions in New York contribute more to the Adirondacks than emissions in Kentucky. Quantitatively, however, the models predict that a unit of emissions in New York contribute from as little as 5 to as much as 30 times as much total suflur deposition in the Adirondacks as a unit of emissions in Kentucky. The practical significance, then, is that we have some confidence that reducing a ton of emissions in New York would yield greater benefit to the Adirondacks thant reducing a ton of emissions from Kentucky. Whether the benefit would be 5 or 30 times as great is uncertain.

Other examples of the variation between models include the following:

(1) **the absolute deposition predicted:** The models predict that the Adirondacks receive from 10 to 34 kg sulfur/hectare/year. For Algoma, the range is 2-14; at Vermont/New Hampshire, the range is 9-25; and at Muskoka, the range is 10-32. The predictions of each model for wet sulfate deposition at various receptors are compared with observed values in Table 6.4; wet deposition was used for this comparison since reliable data is not available for dry deposition. The sulfur deposition predicted by the models was converted to sulfate for this comparison since the observed values were expressed as sulfate.

(2) **the absolute contribution of a given source:** There is significant variation in the absolute deposition predicted to result from given sources. As an example of an extreme, the models predict that the contribution of Minnesota to the Adirondacks varies from 1.3 to 118.8 grams per hectare (a factor of about 90). For about half the source areas, however, the models agree within a factor of ten.

(3) **the percent contribution of a given source:** There is less variation in the quantity than in the absolute contribution described above. With few exceptions, the models agree within a factor of ten on the basis of percentage contribution to the Adirondacks. Among the major contributors, the variation is on the order of 2 or 3; for example, estimates for New York vary from 12-31%, Ohio from 9-17%, Pennsylvania from 10-17%, and Canada from 13-30%. The contributions from states with comparable emissions decreases markedly with increased distance from the receptor; for example, for Kentucky estimates vary from 0.7-2.40% and for Tennessee from 0.0-1.0%.

TABLE 6.4

PREDICTED AND OBSERVED

CURRENT WET SULFATE DEPOSITION

(kg/ha Wet Sulfate)

RECEPTOR	OBSERVED		PREDICTED										
			AES		ASTRAP		ENAMAP		MCARLO		MEP		OME		UMACID	
	O	B	P	E	P	E	P	E	P	E	P	E	P	E	P	E
Adirondacks	31	6	48	17	57	26	25	-6	38	7	18	-13	23	-8	17	-14
Vermont/NH	29	6	41	12	45	16	23	-6	26	-3	17	-12	22	-7	11	-18
Algoma	16	5	28	12	26	10	8	-8	19	3	13	-3	13	-3	8	-8
Muskoka	33	6	46	13	44	11	24	-9	37	4	18	-15	20	-13	23	-10
Average Error			-	14	-	16	-	-7	-	3	-	-11	-	-8	-	-13

O = Observed Measured Value[18]
B = Background[18]
P = Value Predicted by Model Plus Background
E = Difference between Observed and Predicted

Sensitive Area Selection

Four receptor areas were used for this study--the Adirondacks, Vermont/New Hampshire, Algoma and Muskoka. These are among the geographic areas believed to be sensitive to acid deposition due to their bedrock geology (U.S.-Canada Memorandum, 1983: 2-2). Environmental damage, such as lake acidification, has been reported in these areas (U.S.-Canada Memorandum, 1983: 1-1-11-11). Much of the support for reducing acid deposition is a result of concern for ecological damages in these areas (Moynihan, 1984). Similarly, much of the support for acid rain control legislation stems from the belief that such legislation will protect these areas. It is of particular interest to evaluate the impact that the legislation would have on these areas. Hence, while it is arguable whether these areas are the only regions of concern, a control strategy which offers no environmental relief for these areas, or a strategy which would provide little relief to these areas incidental to greater deposition reductions occurring in regions of less concern, would seem misguided.

EMISSION REDUCTION STRATEGIES EVALUATED

Legislative Proposals

The following section describes the sulfur dioxide emission reduction provisions of the major acid rain control bills introduced in Congress. Table 6.3 shows the percentage reduction in sulfur dioxide emissions required of each state under each bill.
1984 Senate Committee Bill--The Senate Environment and Public Works Committee passed comprehensive Clean Air Act amendments in March 1984 which included acid rain provisions. These acid rain provisions call for a 10 million ton reduction in sulfur dioxide by 1995 distributed among the 31 eastern states on the basis of each state's **utility** emissions in excess of 1.2 pounds per million Btu (lbs./mmBtu). 10 million tons constitutes about 44.4% of the 1980 total sulfur dioxide emissions from these states. In general, such an allocation would require large reductions from midwestern states and relatively little reduction from most northeastern states. After 1995, any new plant would have to either meet the "least demonstrated emission limit" or offset the new emissions. It has been predicted that under the bill after 1995, emissions will increase above 1995 levels as a result of new plants meeting these limits (ICF, Inc., 1983:

9).
　　Sikorski/Waxman Bill--This bill would require a 10 million ton reduction distributed among 48 states based on **utility** emissions in excess of 1.2 pounds per million Btu.
　　D'Amours Bill--Also referred to as the New England Caucus Bill, this bill would require a 12 million ton reduction in the 48 states. The method of allocating emissions reductions incorporates some (although not proportional) recognition that non-utility sources are significant emitters of sulfur dioxide; non-utility sources are responsible for about one-third of the sulfur dioxide emissions in the 48 state region and about half of the sulfur dioxide emissions in the Northeast. Of the 12 million ton reduction target, 10 million would be allocated based on **utility** emissions in excess of 1.2 lbs/mmBtu, 1 million would be allocated based on non-utility boilers in excess of 1.2 lbs./mmBtu, and 1 million would be allocated based on industrial processes emitting in excess of the national average Best Available Control Technology.
　　Stafford Bill--This bill would require a 12 million ton reduction in sulfur dioxide emissions from the 31 eastern states (or about a 53% overall reduction). Emission reductions would be allocated to the individual states on the basis of **utility** emissions in excess of 1.2 pounds of SO_2 per million Btu.

Uniform Percentage Reduction Strategy

　　This hypothetical strategy would entail a 10 million ton reduction in the thirty-one eastern states as would the Senate Committee Bill. It differs from the Senate Committee Bill in that it allocates reductions to the states based on each state's **total** sulfur dioxide emissions (as opposed to **utility** emissions in excess of 1.2 lbs/mmBtu). Such an allocation would result in a requirement for each state to reduce by an equal percentage of about 44.4%. (States with higher emissions would, of course, be required to reduce by a higher tonnage than states with lower emissions.)
　　Two general observations about this strategy can be made. First, it would require each state to reduce in proportion to how much it emits. Although it has been stated that the legislative proposals would require the largest reductions from states with the greatest emissions (Moynihan, 1984), this statement is not true. As an example, in 1980 New York and Kentucky each emitted about the same amount of sulfur dioxide; under the Senate Committee Bill, New York would be required to reduce emissions by about 21% but Kentucky by 63%. This is due largely to the use of **utility**

emissions for assigning reduction requirements. A second observation is that we may have greater confidence in the results produced by a uniform reduction strategy. Under a "true" uniform reduction strategy (where **every** source reduces by an equal percentage) every model will predict the same result--that deposition will decrease by that same percentage. This is so because, if each state's emissions are reduced by a certain percentage, each state's contribution to the deposition will be reduced by the certain percentage, and therefore, the total deposition will be reduced by that same percentage. This approach is in keeping with a statement by the National Academy of Sciences: "It is the Committee's judgment that if the emissions of sulfur dioxide from all sources were reduced **by the same fraction,** the result would be a corresponding fractional reduction in deposition" (emphasis added) (NRC, 1983). By comparison, under non-uniform strategies (i.e., legislative proposals), each model predicts a different result because each model predicts a different relative contribution from each state. (NOTE: The strategy presented here is not a "true" uniform reduction since emissions from the 17 western states and Canadian sources are not reduced. Every model will predict the same decrease (44.4%) in the contribution from the thirty-one states. The total decrease in deposition predicted will vary between models since each model predicts a different contribution from these other sources.)

Elimination of Power Plant SO_2

A more extreme emission reduction strategy was evaluated--complete elimination of SO_2 from all electrical-generating power plants in the 31 states area and Canada. This strategy was selected because of the focus on electric utilities for emission reduction allocation common to all legislative proposals. Electric utility sulfur dioxide emissions are a major factor in the acid rain equation. Recognition of this has led to control proposals that ignore all other contributors--with the implication that reduction of utility sulfur dioxide emissions alone will lead to acceptable control of acid rain. This analysis was conducted to examine the effects of extreme controls on electric utility sulfur dioxide emissions. This strategy would entail a reduction in emissions of about 0.8 million tons from Canada and over 16 million from the U.S. or about 17 million tons total. It is essentially a 17 million ton reduction requirement allocated to the eastern U.S. and Canada based on each "state's" **utility** emissions.

Illustrative Alternatives

Four alternatives are presented for illustrative purposes only--specifically to illustrate to what extent a control strategy designed to achieve specific results will be more efficient than strategies which achieve these results incidentally and to illustrate how the transfer matrices can be used to design control strategies.

All four strategies were designed to produce the same reduction in total sulfur deposition as the Senate Committee Bill at the four receptors according to the MCARLO model. They were designed using a technique of linear programming. The technique was used to design strategies which would produce the same reduction in sulfur deposition at the four receptors as the Senate Committee Bill while minimizing the emission reductions necessary to achieve these results. This procedure then represents the normal approach of addressing environmental problems. First, the desired environmental changes are specified and then a control strategy to achieve these changes is designed. The reductions predicted to occur under the Senate Committee Bill were used as the desired changes to permit comparison.

Alternative 1 involves emission reductions achieved only in the thirty-one eastern states as does the Senate Committee Bill. Alternative 2 is identical except that no state would reduce emissions by more than 70%. Alternative 3 involves emission reductions in both the U.S. and Canada. Alternative 4 is identical to alternative 3 except that no "state" would reduce by more than 70%. Alternatives 1 and 2 can be more directly compared to the Senate Committee bill in that emission reductions are limited to the same geographic area. Alternatives 3 and 4 are included to illustrate how an international strategy may be designed.

RESULTS AND COMPARISON OF CONTROL STRATEGIES

The complete results of each control strategy estimated by each model on the four receptor sites are tabulated in Appendix A. Direct comparison of all strategies is difficult because different control strategies entail different total emission reductions and because each of the seven models predicts a different benefit under a given emission reduction strategy. Condensed results are shown in Tables 6.5a-d and 6.6a-d. These condensed results show the range predicted by seven models and the results predicted by one model, MCARLO. (Although it is beyond the scope of

TABLE 6.5a

PERCENT REDUCTION IN SULFUR DEPOSITION AND
EFFICIENCY OF VARIOUS CONTROL STRATEGIES

RECEPTOR: Adirondacks

Emission Reduction Scenario	Efficiency[a]		% Reduction	
	RANGE	MCARLO	RANGE	MCARLO
CURRENT	-	-	-	-
SENATE BILL	233-1071	1071	24.1-32.8	31.3
SIKORSKI/WAXMAN	237-1076	1076	28.6-33.6	32.6
D'AMOURS	232-1043	1043	32.3-38.4	36.9
STAFFORD	234-1073	1073	29.2-39.5	37.7
UNIFORM	307-1229	1229	30.6-38.5	35.8
ELIM. POWER PLANT SO$_2$	265-1130	1130	46.1-57.3	55.5
ALTERNATIVE 1	577-2380	2176	27.5-39.1	31.3
ALTERNATIVE 2	450-1937	1785	27.2-38.8	31.3
ALTERNATIVE 3	1148-4131	3550	27.5-41.6	31.3
ALTERNATIVE 4	955-3016	2992	27.2-39.2	31.3

[a]Efficiency is in units of grams of sulfur per hectare deposition reduction per million tons of sulfur dioxide emissions reduced

TABLE 6.5b

PERCENT REDUCTION IN SULFUR DEPOSITION AND
EFFICIENCY OF VARIOUS CONTROL STRATEGIES

RECEPTOR: Vermont/New Hampshire

Emission Reduction Scenario	Efficiency[a]		% Reduction	
	RANGE	MCARLO	RANGE	MCARLO
CURRENT	-	-	-	-
SENATE BILL	174-666	627	18.0-29.1	26.0
SIKORSKI/WAXMAN	181-676	636	19.3-30.6	27.3
D'AMOURS	178-652	620	23.2-34.4	31.1
STAFFORD	175-668	629	21.7-35.0	31.3
UNIFORM	248-776	748	23.1-33.8	30.9
ELIM. POWER PLANT SO_2	205-696	677	35.6-51.1	47.2
ALTERNATIVE 1	390-1499	1275	20.0-32.1	25.9
ALTERNATIVE 2	310-1224	1045	19.1-32.0	26.0
ALTERNATIVE 3	700-2681	2381	24.2-35.2	29.7
ALTERNATIVE 4	726-2700	2450	30.7-42.2	36.3

[a]efficiency is in units of grams of sulfur per hectare deposition reduction per million tons of sulfur dioxide emissions reduced

TABLE 6.5c

PERCENT REDUCTION IN SULFUR DEPOSITION AND
EFFICIENCY OF VARIOUS CONTROL STRATEGIES

RECEPTOR: Algoma

Emission Reduction Scenario	Efficiency[a] RANGE	Efficiency[a] MCARLO	% Reduction RANGE	% Reduction MCARLO
CURRENT	-	-	-	-
SENATE BILL	47-433	341	21.4-37.1	28.5
SIKORSKI/WAXMAN	46-432	344	22.0-38.4	29.8
D'AMOURS	40-386	284	21.8-40.0	28.7
STAFFORD	47-433	342	25.7-44.6	34.3
UNIFORM	46-436	337	25.5-37.3	28.1
ELIM. POWER PLANT SO_2	44-419	333	33.5-60.4	46.8
ALTERNATIVE 1	89-934	695	20.0-39.2	28.5
ALTERNATIVE 2	87-826	705	23.7-42.4	35.3
ALTERNATIVE 3	274-1251	1130	14.0-42.2	28.5
ALTERNATIVE 4	227-1108	958	25.3-39.5	28.6

[a]efficiency is in units of grams of sulfur per hectare deposition reduction per million tons of sulfur dioxide emissions reduced

TABLE 6.5d

PERCENT REDUCTION IN SULFUR DEPOSITION AND
EFFICIENCY OF VARIOUS CONTROL STRATEGIES

RECEPTOR: Muskoka

Emission Reduction Scenario	Efficiency[a]		% Reduction	
	RANGE	MCARLO	RANGE	MCARLO
CURRENT	-	-	-	-
SENATE BILL	275-888	888	14.7-33.9	27.5
SIKORSKI/WAXMAN	275-879	879	15.0-34.8	28.3
D'AMOURS	255-838	838	17.4-38.1	31.5
STAFFORD	276-888	888	17.7-40.7	33.1
UNIFORM	280-856	856	14.6-32.4	26.5
ELIM. POWER PLANT SO$_2$	301-967	967	34.0-61.3	50.5
ALTERNATIVE 1	563-1802	1802	18.7-30.6	27.5
ALTERNATIVE 2	473-1478	1478	16.3-33.3	27.5
ALTERNATIVE 3	1156-4365	4060	26.1-64.7	37.9
ALTERNATIVE 4	960-3164	3164	30.4-53.8	35.2

[a]Efficiency is in units of grams of sulfur per hectare deposition reduction per million tons of sulfur dioxide emissions reduced

TABLE 6.6a

pH OF RAINFALL PREDICTED TO
OCCUR UNDER CONTROL STRATEGIES

RECEPTOR: Adirondacks

Emission Reduction Scenario	pH	
	RANGE	MCARLO
CURRENT	4.2	4.2
SENATE BILL	4.29-4.34	4.32
SIKORSKI/WAXMAN	4.29-4.34	4.32
D'AMOURS	4.31-4.36	4.34
STAFFORD	4.31-4.37	4.34
UNIFORM	4.30-4.34	4.32
ELIM. POWER PLANT SO_2	4.38-4.45	4.42
ALTERNATIVE 1	4.29-4.35	4.29
ALTERNATIVE 2	4.29-4.33	4.29
ALTERNATIVE 3	4.25-4.33	4.27
ALTERNATIVE 4	4.26-4.31	4.28

TABLE 6.6b

pH OF RAINFALL PREDICTED TO OCCUR UNDER CONTROL STRATEGIES

RECEPTOR: Vermont/New Hampshire

Emission Reduction Scenario	pH RANGE	MCARLO
CURRENT	4.26	4.26
SENATE BILL	4.33-4.37	4.36
SIKORSKI/WAXMAN	4.33-4.38	4.36
D'AMOURS	4.34-4.39	4.38
STAFFORD	4.35-4.40	4.38
UNIFORM	4.34-4.39	4.37
ELIM. POWER PLANT SO_2	4.40-4.48	4.45
ALTERNATIVE 1	4.33-4.38	4.34
ALTERNATIVE 2	4.33-4.37	4.34
ALTERNATIVE 3	4.32-4.38	4.34
ALTERNATIVE 4	4.34-4.40	4.35

TABLE 6.6c

pH OF RAINFALL PREDICTED TO
OCCUR UNDER CONTROL STRATEGIES

RECEPTOR: Algoma

Emission Reduction Scenario	pH RANGE	MCARLO
CURRENT	4.35	4.35
SENATE BILL	4.40-4.48	4.46
SIKORSKI/WAXMAN	4.40-4.49	4.47
D'AMOURS	4.40-4.50	4.47
STAFFORD	4.41-4.52	4.49
UNIFORM	4.40-4.48	4.46
ELIM. POWER PLANT SO_2	4.43-4.59	4.55
ALTERNATIVE 1	4.41-4.50	4.44
ALTERNATIVE 2	4.41-4.50	4.48
ALTERNATIVE 3	4.38-4.58	4.41
ALTERNATIVE 4	4.41-4.54	4.42

TABLE 6.6d

pH OF RAINFALL PREDICTED TO OCCUR UNDER CONTROL STRATEGIES

RECEPTOR: Muskoka

Emission Reduction Scenario	pH RANGE	MCARLO
CURRENT	4.2	4.2
SENATE BILL	4.27-4.32	4.31
SIKORSKI/WAXMAN	4.27-4.33	4.32
D'AMOURS	4.28-4.34	4.33
STAFFORD	4.28-4.35	4.34
UNIFORM	4.26-4.31	4.30
ELIM. POWER PLANT SO_2	4.34-4.44	4.41
ALTERNATIVE 1	4.28-4.31	4.30
ALTERNATIVE 2	4.27-4.31	4.30
ALTERNATIVE 3	4.27-4.39	4.28
ALTERNATIVE 4	4.28-4.36	4.28

this analysis to select a "best" model, the MCARLO model predicts current wet deposition with the least overall error (see Table 6.4) predicts results intermediate of the seven models, and has been suggested by Dr. James Young and others as the model which performs best.) Table 6.5 shows the percent reduction in total sulfur deposition and the efficiency (described below). Table 6.6 shows the resulting pH of rainfall as calculated from the reduction in wet sulfur deposition.

(a) **pH**

The decrease in acidity resulting from large sulfur dioxide emission reduction is less than many would expect. For example, the 12 million ton reduction is predicted to increase pH at the Adirondacks from the current 4.2 to 4.37 according to the most optimistic model. Total elimination of sulfur dioxide from all power plants is predicted to increase pH at the Adirondacks to only 4.38-4.45. These increased pH values fall short of the levels of 4.6-4.7 mentioned by the National Academy of Sciences as desirable (NRC, 1981: 181). They fall far short of the call to "stop acid rain."

(b) **Efficiency**

To normalize the results for the different quantities of total emission reductions involved in different control strategies, the change in deposition is divided by the emission reduction requirement of each strategy to represent efficiency of the control strategy. This value is in units of grams of reduced sulfur deposition per hectare per million tons of sulfur dioxide emission reduced--in other words, how much benefit for the average ton of emission reduced. The matrix coefficients themselves serve as a measure of efficiency; the overall efficiencies are essentially an average of the coefficients, weighted by the magnitude of emission reduction in each source area.

General observations can be made about the efficiency of the various strategies. The illustrative alternatives are significantly more efficient than any other strategy considered. The total emissions reductions required under alternatives 1, 2, 3, and 4 are 4.9, 6.0, 3.0, and 3.6 million tons, respectively. The reductions required of individual states are compared with those required under the Senate Bill and shown as Table 6.7. Since these alternatives produce the same results at the four receptors as the Senate Bill (which would entail 10 million tons reduction)

TABLE 6.7

PERCENT REDUCTION IN SULFUR DIOXIDE (SO_2) EMISSIONS UNDER
VARIOUS CONTROL STRATEGIES
FOR THE UNITED STATES

State	Senate Bill	Alt 1	Alt 2	Alt 3	Alt 4
Alabama	41.0	0.0	0.0	0.0	0.0
Arkansas	0.7	0.0	0.0	0.0	0.0
Connecticut	0.0	0.0	0.0	0.0	0.0
Delaware	18.7	0.0	0.0	0.0	0.0
D.C.	0.0	0.0	0.0	0.0	0.0
Florida	36.5	0.0	0.0	0.0	0.0
Georgia	54.0	0.0	0.0	0.0	0.0
Illinois	50.2	0.0	70.0	0.0	0.0
Indiana	57.3	0.0	25.0	0.0	0.0
Iowa	37.4	68.0	70.0	0.0	3.5
Kentucky	62.9	0.0	0.0	0.0	0.0
Louisiana	0.0	0.0	0.0	0.0	0.0
Maine	2.8	0.0	0.0	0.0	0.0
Maryland	33.3	0.0	0.0	0.0	0.0
Massachusetts	22.2	0.0	0.0	0.0	0.0
Michigan	25.6	100.0	70.0	15.0	70.0
Minnesota	22.8	0.0	0.0	0.0	0.0
Mississippi	37.3	0.0	0.0	0.0	0.0
Missouri	67.6	0.0	0.0	0.0	0.0
N. Hampshire	46.1	56.0	47.0	100.0	70.0
New Jersey	10.6	0.0	0.0	0.0	0.0
New York	21.0	81.0	70.0	100.0	70.0
N. Carolina	14.4	0.0	0.0	0.0	0.0
Ohio	57.8	85.0	70.0	0.0	0.0
Pennsylvania	40.5	0.0	14.0	0.0	22.0
Rhode Island	0.0	0.0	0.0	0.0	0.0
S. Carolina	30.7	0.0	0.0	0.0	0.0
Tennessee	62.5	0.0	0.0	0.0	0.0
Vermont	0.7	56.0	47.0	100.0	70.0
Virginia	8.0	0.0	0.0	0.0	0.0
W. Virginia	53.2	0.0	0.0	0.0	0.0
Wisconsin	49.8	68.0	70.0	0.0	3.5
(West)*					
W.N.E.	0.0	0.0	0.0	0.0	0.0
W.S.E.	0.0	0.0	0.0	0.0	0.0
W.N.W.	0.0	0.0	0.0	0.0	0.0
W.S.W.	0.0	0.0	0.0	0.0	0.0

*Western regions:
W.N.E. includes Nebraska, North Dakota, South Dakota, Montana, and Wyoming
W.S.E. includes Oklahoma, Kansas, Colorado, New Mexico, and Texas
W.N.W. includes Washington, Idaho, and Oregon
W.S.W. includes California, Nevada, Utah, and Arizona

TABLE 6.7 (continued)

PERCENT REDUCTION IN SULFUR DIOXIDE (SO_2)
EMISSIONS FOR CANADIAN SOURCES

Canadian Sources	Alt 1	Alt 2	Alt 3	Alt 4
Northern Manitoba	0.0	0.0	0.0	0.0
Southern Manitoba	0.0	0.0	0.0	0.0
Northwestern Ontario	0.0	0.0	0.0	0.0
Northeastern Ontario	0.0	0.0	100.0	70.0
Sudbury, Ontario	0.0	0.0	100.0	70.0
Southwestern Ontario	0.0	0.0	88.0	70.0
Southeastern Ontario	0.0	0.0	100.0	70.0
Montreal & St. Lawrence Valley	0.0	0.0	0.0	70.0
Norando & North Central Quebec	0.0	0.0	0.0	0.0
Gaspe Bay, Quebec	0.0	0.0	0.0	0.0
New Brunswick	0.0	0.0	0.0	0.0
Nova Scotia & Prince Edward Island	0.0	0.0	0.0	0.0
Newfoundland & Labrador	0.0	0.0	0.0	0.0
Saskatchewan & Alberta	0.0	0.0	0.0	0.0
British Columbia & Yukon	0.0	0.0	0.0	0.0

with much less emission reduction, it is apparent that they are much more efficient. Alternative 3, for example, is over three times as efficient in that less than one-third as much emission reduction is necessary.
It is noted that a strategy which produces the same results as the Senate Committee Bill at four receptors will not necessarily produce the same results everywhere; the only strategy which would do so would, of course, be identical to the Senate Committee Bill. The results produced by the Senate Committee Bill are not necessarily what is desired. Many people mistakenly believe that legislative proposals are designed to reduce deposition in the sensitive areas. They are not. Under the Senate Committee Bill, the greatest reduction in deposition has been predicted to occur in the Upper Ohio River Valley (Streets, et al., 1983: 474A-485A); many people believe the greatest deposition reduction should occur in the northeastern U.S. (or other areas where damage is reported or feared).
The alternatives were designed using an optimization technique which determines the least amount of emission reduction necessary to achieve the specified goals; they are designed to reduce deposition in the geographic areas specified.
The uniform reduction strategy offers slightly greater efficiency in the Adirondacks and New England than legislative proposals because it allocates reductions based on **total** emissions rather than power plant emissions; allocation by power plant emissions ignores important non-utility sources, many of which are located closer to the Adirondacks and New England. In the Northeast, approximately half the sulfur dioxide emissions are non-utility. Thus, allocation of emission reductions based on utility emissions (as opposed to total emissions) creates a bias toward less effective emission reductions, i.e., reductions would be achieved in the wrong places.

IMPORTANCE OF EMISSION GROWTH

This chapter has analyzed the reduction in deposition that may result from reducing emissions below 1980 levels by the amounts specified by legislative proposals (and hypothetical control strategies). In order to maintain the reduced level of deposition, it would be necessary to maintain the reduced level of emissions. In practice, this would mean that any "new" emissions (resulting from new plants or increased utilization of existing plants) must be negated or offset by further reductions at existing plants. The economic implications are severe and have been documented by others (Enoch, 1982: 29-35).

It is somewhat unclear to what extent the legislative proposals would require offsetting. This chapter analyzes emissions when they are at the lowest level and, in some cases, lower than they ever might be under a particular control strategy. (Under the Waxman/Sikorski and D'Amours bills, for example, growth without offsets appears to be allowed; thus emissions would never reach a level of 10 or 12 million tons below 1980 levels.) Obviously, if emissions are allowed to increase from the reduced level, the benefit of the reduction will be diminished; the results shown here may be viewed as representing "best case" conditions. As an example, a projection of emissions under a bill passed by the Senate committee in 1982 shows emissions being reduced by the required 8 million tons such that in the year 1995, emissions are 8 million tons below 1980 levels. However, since the legislation appears to allow new sources meeting the "least demonstrated emission limit" to be built without offsetting their new emissions, it was predicted that emissions would increase above the 1995 reduced level. Interestingly, it was predicted that under the current Clean Air Act, emissions will begin to decrease in the year 1995 such that at some point, emissions will be virtually the same with or without the legislation (ICF, Inc., 1983: 9).

Additionally, for this analysis, emissions of nitrogen oxides were assumed to remain constant at 1980 levels. In fact, they have been predicted to increase (NRC, 1981: 47). If they do increase, the improvements in rainfall pH presented here are overstated.

EFFECT OF NATURAL DEPOSITION

This chapter examines only man-made emissions and deposition of sulfur and nitrogen oxides although it is widely accepted that some portion of current deposition is due to biogenic and other natural sources of emissions. Thus, the benefits presented herein are again overstated to some degree.

The 1983 report of the National Acid Precipitation Assessment Program found that recent studies suggest that the unpolluted pH of precipitation is closer to 5.0 (Interagency Task Force, 1983: 13). One estimate, for example, is that natural wet sulfate deposition may be equal to 10 kg/ha (Gorham, et al., 1984: 408). Thus, at the Adirondacks, where wet deposition has been estimated to be 31 kg/ha, if the man-made contribution to wet deposition is reduced by 35% (as predicted by the MCARLO model under the Senate Bill), total deposition would be reduced by only 24%. Similarly, although this analysis assumes that the sulfuric acid component of rainfall would decrease by 35% resulting in a pH change from the current 4.2 to 4.32, a 24% reduction may be more realistic which would increase pH to only 4.28. Additionally, the contribution of atmospheric carbon dioxide is ignored in this study although it is generally accepted that this contribution would theoretically cause "natural" rainfall to have a pH of 5.6. However, in the pH ranges under consideration, incorporating this contribution would not make a significant difference.

IMPLICATIONS FOR CONTROL STRATEGY

The matrix coefficients enable one to determine **where** a unit of emission reduction would provide the greatest benefit to a given receptor. For example, if one sought to reduce deposition in the Adirondacks, there is general agreement by all the models that emission reductions in New York and Southeastern Ontario would produce the greatest benefit per unit of emission reduction. By comparison, emission reductions achieved in the Midwest, at coal-fired power plants for example, would produce much less benefit. For example, the models indicate that reductions in New York would produce 5-30 times as much benefit as equivalent reductions in Kentucky.

One method of designing a control strategy could incorporate such scientific information; to date, legislative proposals have determined where (and how much) emissions would be reduced based on political judgment. This chapter shows that such proposals are not nearly as effective as a strategy designed to achieve specific results; in fact, in some cases, such a basis actually creates a bias towards inefficiency.

The matrix coefficients can be used directly to estimate how best to achieve desired environmental changes while entailing the least emission reduction as was done in the illustrative alternatives. This would not necessarily be the least-cost strategy since the cost of reducing a unit of emission differs from plant to plant. Further, at a given plant, the cost per ton of emission reduction generally increases as greater reductions are sought; i.e., the marginal cost increases. Design of a least-cost strategy would incorporate the difference in the marginal costs of reducing emissions from different sources.

The allocation of required emission reductions under current legislative proposals has been defended as representing close to the least-cost method of achieving a given target emission reduction. In other words, the cost of achieving the last ton of required emission reduction is virtually the same in each state. What this actually means in terms of cost-effectiveness is that the same amount of money would be paid for reducing a ton of emissions in Kentucky as would be paid to reduce a ton of emissions in New York although all the models agree that much greater benefit would result at the Adirondacks and New England from the reduction achieved in New York. Thus, it would be more cost-effective to take the money which would be spent reducing the ton of emissions in Kentucky and use it to reduce an additional ton in New York. For total sulfur deposition in the Adirondacks, the MCARLO model indicates that the ton of emission reduction in New York would produce over nine times the benefit of reducing the ton in Kentucky. Hence, it would be more cost-effective to reduce emissions in New York until the marginal cost of reducing emissions became over nine times that of reducing emissions in Kentucky, according to this model. One model (MEP) predicts that the benefit of reducing the ton in New York would produce 30 times the benefit of reducing the ton in Kentucky. Hence, for the same amount of money, up to 30 times the benefit could be realized.

Once the relationship between the benefits (measured as estimated reduction in deposition, for example) and the marginal costs of reductions in different areas are quantified, a cost-effective control strategy can be designed. Various constaints can be incorporated. For example, desired environmental benefit can be made the controlling factor by specifying what level of deposition is sought and designing the strategy to achieve it. (The illustrative alternatives presented in this chapter used specified deposition levels as the controlling factor.) Alternatively, a limit on the cost of the control program may be established and a strategy designed to yield the greatest benefit for this set cost may be designed. Another alternative might be to place a limit on the cost-effectiveness. In other words, specify that the most cost-effective reductions should continue until the next increment of benefit would cost more than a specified amount. Or, specify how much it is worth to achieve an additional unit of benefit. This approach represents traditional cost-benefit analysis which is used in addressing environmental and other problems. It would seem to be much more appropriate than the approach used in formulating legislative proposals.

CONCLUSIONS

One conclusion which stands out from this analysis is that it is inappropriate to compare directly estimates of the costs of damage due to acid rain with the estimated costs of legislative control programs since a small decrease in rainfall acidity is predicted from legislative proposals. Even such drastic measures as total elimination of sulfur dioxide emissions from all power plants is predicted to result in a small decrease in rainfall acidity--even by the most "optimistic" model. In turn, the environmental benefits, in terms of reduced risk of aquatic ecosystems in sensitive areas for example, will be much less than expected by many. It is fairly certain that whatever costs are associated with current and future environmental damage due to acid rain will not be eliminated by legislative proposal or even more drastic measures.

Secondly, all the models indicate that the expected benefit from reducing emissions is highly dependent on where the reduction is achieved. Legislative proposals do not take this fact into account except for limiting the reductions to 31 states as opposed to 48 (or 50); distribution of emission reductions within the specified region is done without any regard for geographic location. In fact, the basis for distributing emission reductions actually causes the reductions to be very ineffective. Although there is often discussion of how large an emission reduction is needed, such discussions are meaningless unless the locations of the emission reductions are considered. The evaluation of the various emission control strategies indicates that the importance of geographic location may be too significant to ignore; the alternative strategies, for example, achieve equivalent benefit at the receptors with significantly less emission reductions than the legislative proposals.

A more efficient control program can be designed by developing the physical characteristics of the strategy independent of equity issues. Equity and other concerns can then be incorporated without sacrificing efficiency. Some have suggested that proposed legislative strategies represent fair and equitable allocations of emission reductions; others contend they do not. This chapter did not delve into the complicated philosophical questions of what is fair and equitable; the results of its evaluation indicate that the legislation proposals are not nearly as efficient as a strategy which defines the desired environmental changes before formulating a solution. One questions the value of a purportedly "fair and equitable" solution that is inefficient. If, in fact, a solution that is based on this perception of fair and equitable allocations yields an inefficient or ineffective solution (and this chapter concludes that the allocation method contained in legislative proposals actually creates a bias towards inefficiency), then perhaps a different approach should be taken.

Perhaps a solution that will be effective and efficient should first be devised without regard to equity; equity, fairness, and other concerns could then be incorporated by who pays for the reductions and how the reductions are achieved. There is growing support for the concept of a nationally funded emission control program which minimizes hardship on any particular group and region; the Sikorski/Waxman bill has over a hundred co-sponsors, for example. Unfortunately, this bill starts with an inefficient emission reduction strategy. Once the physical characteristics of an effective program are devised, these other concepts may be incorporated through some form of national funding mechanism.

ADDENDUM: A NOTE ON THE WORKS OF OTHERS

Researchers at Argonne National Laboratory published results of their analyses of several legislative proposals (Streets, et al., 1983). Their analyses, using a more sophisticated application of the ASTRAP model, predicted that the reduction in total (wet and dry) sulfur deposition at the Adirondacks would be 29% and 37% under the Senate committee bill (S.3041) and the D'Amours bill (H.R. 4404) respectively. Assuming that this would reduce the sulfuric acid component of rainfall accordingly (assuming that wet and dry deposition are approximately equal), this would result in an increase of pH from 4.2 to 4.3 and 4.32 respectively--comparable with the findings of this chapter. The authors noted that the greatest reduction in deposition occurred in the Ohio River Basin area. The authors wrote: "The deposition reductions in areas with sensitive ecosystems such as the Adirondacks and Great Smoky Mountains are significantly less, drawing into question the effectiveness of such measures."

Researchers at the MIT Energy Laboratory have published the results of calculating the expected change in the average pH of rainfall in the northeastern quadrant of the U.S. (19 states) (Fay, et al., 1983: 17). These researchers predicted that the Senate Committee Bill would increase the average pH from the current 4.2 to 4.3 (at the time of their publication, this bill was referred to as the Mitchell

Bill). In addition, several hypothetical control strategies were analyzed. The results are informative and, in the words of the researchers, probably discouraging. For example, complete elimination of all sulfur dioxide from all power plants was predicted to result in an increase in the pH to only 4.4--again comparable to the findings of this chapter. Complete elimination of all sulfur and nitrogen emissions from all power plants and industrial boilers was predicted to increase the average pH to 4.7.

REFERENCES

Enoch, Harry C. **Economic Impacts of the Senate Acid Rain Bill: Some Implications for Kentucky.** Lexington, Ky.: Kentucky Energy Cabinet, December 1982.

Fay, James A., Dan Golumb, and James Gruhl. "Controlling Acid Rain." Cambridge, Mass.: M.I.T. Energy Laboratory, April 1983.

Gorham, Eville, Frank B. Martin, and Jack T. Litzau. "Acid Rain: Ionic Correlations in the Eastern United States, 1980-1981." Science. 225 (July 1984).

ICF, Inc. "Analysis of a Senate Emission Reduction Bill (S.3041). Washington: U.S. Environmental Protection Agency, February 1983.

Interagency Task Force on Acid Precipitation, National Acid Precipitation Assessment Program. **Annual Report 1983 to the President and Congress.** Washington, D.C.: U.S.G.P.O. 1983.

Moynihan, Daniel Patrick. Remarks before the Acid Rain Policy Forum, Albany, New York, February 8, 1984.

National Research Council, National Academy of Sciences. **Atmosphere-Biosphere Interactions: Toward a Better Understanding of the Ecological Consequences of Fossil Fuel Combustion.** Washington, D.C.: National Academy Press, 1981.

National Research Council, National Academy of Sciences. **Acid Deposition: Atmospheric Processes in Eastern North America.** Washington, D.C.: National Academy Press, 1983.

Streets, David G., Duane A. Knudsen, and Jack Shannon. "Selected Strategies to Reduce Acid Deposition in the U.S." **Environmental Science & Technology.** 17 (1983) 474-485.

U.S.-Canada Memorandum of Intent on Transboundary Air Pollution, Work Group 1. **Impact Assessment: Final Report.** January 1983.

U.S.-Canada Memorandum of Intent on Transboundary Air Pollution, Work Group 2. **Atmospheric Sciences and**

Analysis. Report No. 2F-M. November 1982.
U.S. Environmental Protection Agency, Acid Deposition Task Force. **Briefing Document for the Administrator. Acid Deposition: Current Knowledge and Policy Options.** Washington, D.C.: U.S. EPA, August 1983.

APPENDIX A

RECEPTOR: Muskoka

Emission Reduction Scenario	Deposition Predicted by Transfer Matrix (kg total sulfur/ha/yr)						
	AES	ASTRAP	ENAMAP	MCARLO	MEP	OME	UMACID
CCURRENT	27.9	20.8	20.3	32.3	11.7	10.3	15.0
SENATE BILL	20.5	13.8	17.3	23.4	8.9	7.5	11.1
(% Reduction)	26.4%	33.9%	14.7%	27.5%	24.1%	26.8%	26.2%
(Efficiency)	737	705	299	888	282	275	393
SIKORSKI/WAXMAN	20.3	13.6	17.3	23.1	8.8	7.4	10.9
(% Reduction)	27.2%	34.8%	15.0%	28.3%	24.9%	27.8%	27.0%
(Efficiency)	731	697	295	879	281	275	389
D'AMOURS	19.5	12.9	16.8	22.1	8.4	7.2	10.5
(% Reduction)	30.3%	38.2%	17.4%	31.5%	28.1%	30.1%	29.8%
(Efficiency)	697	656	292	838	271	255	369
STAFFORD	19.0	12.3	16.7	21.6	8.3	7.0	10.3
(% Reduction)	31.8%	40.7%	17.7%	33.1%	29.0%	32.3%	31.6%
(Efficiency)	738	705	299	888	282	276	393
UNIFORM	20.7	14.1	17.4	23.7	8.9	7.5	11.1
(% Reduction)	25.9%	32.4%	14.6%	26.5%	24.0%	27.5%	25.9%
(Efficiency)	724	675	297	856	280	284	388
ELIM. POWER PLANT SO$_2$	14.6	8.1	13.4	16.0	5.4	5.2	7.1
(% Reduction)	47.6%	61.3%	34.0%	50.5%	54.2%	49.3%	52.4%
(Efficiency)	788	757	409	867	376	301	466
ALTERNATIVE 1	20.4	14.4	16.5	23.4	8.6	7.5	11.0
(% Reduction)	26.9%	30.6%	18.7%	27.5%	26.5%	26.9%	26.5%
(Efficiency)	1525	1297	775	1802	630	563	809
ALTERNATIVE 2	20.4	13.9	17.0	23.4	8.7	7.5	10.9
(% Reduction)	27.1%	33.3%	16.3%	27.5%	25.5%	27.6%	27.0%
(Efficiency)	1261	1154	553	1478	496	473	674
ALTERNATIVE 3	16.7	15.4	7.2	20.0	6.5	6.8	8.7
(% Reduction)	40.0%	26.1%	64.7%	37.9	44.5	33.8	41.6
(Efficiency)	3705	1801	4365	4060	1728	1156	2072
ALTERNATIVE 4	17.5	14.5	9.4	20.9	6.9	6.8	9.2
(% Reduction)	37.2%	30.4%	53.8%	35.2%	41.1%	33.4%	38.9%
(Efficiency)	2890	1765	3050	3164	1341	960	1624

NOTE: Efficiency is in units of reduced deposition (in grams of sulfur per hectare per year) per reduced emissions (million tons of sulfur dioxide)

APPENDIX A

RECEPTOR: Algoma

Emission Reduction Scenario	Deposition Predicted by Transfer Matrix (kg total sulfur/ha/yr)						
	AES	ASTRAP	ENAMAP	MCARLO	MEP	OME	UMACID
CURRENT	13.6	11.7	2.2	12.0	4.4	6.2	3.9
SENATE BILL	9.5	7.4	1.7	8.6	3.4	4.4	2.6
(% Reduction)	30.3%	31.1%	21.4%	28.5%	23.4%	29.7%	34.1%
(Efficiency)	413	433	47	341	103	185	134
SIKORSKI/WAXMAN	9.4	7.2	1.7	8.4	3.3	4.3	2.5
(% Reduction)	31.4%	38.4%	22.0%	29.8%	24.2%	31.2%	35.3%
(Efficiency)	411	432	46	344	102	187	134
D'AMOURS	9.3	7.0	1.7	8.6	3.3	4.2	2.5
(% Reduction)	32.0%	40.0%	21.8%	28.7%	24.2%	32.3%	35.7%
(Efficiency)	359	386	40	284	88	166	116
STAFFORD	8.6	6.5	1.6	7.9	3.2	4.0	2.3
(% Reduction)	36.5%	44.6%	25.7%	34.3%	28.2%	35.9%	41.0%
(Efficiency)	413	433	47	342	103	186	134
UNIFORM	9.5	7.3	1.7	8.6	3.3	4.3	2.6
(% Reduction)	30.4%	37.3%	20.9%	28.1%	25.5%	31.3%	34.8%
(Efficiency)	414	436	46	337	112	195	137
ELIM. POWER PLANT SO$_2$	6.8	4.6	1.5	6.4	2.5	3.0	1.7
(% Reduction)	50.3%	60.4%	33.5%	46.8%	42.4%	51.5%	57.8%
(Efficiency)	406	419	44	333	111	191	135
ALTERNATIVE 1	9.8	7.1	1.7	8.6	3.0	4.3	2.4
(% Reduction)	28.1%	39.2%	20.0%	28.5%	31.9%	30.7%	38.8%
(Efficiency)	779	934	89	695	286	390	311
ALTERNATIVE 2	9.0	6.7	1.7	7.8	3.0	4.3	2.3
(% Reduction)	34.1%	42.4%	23.7%	35.3%	32.9%	31.8%	41.1%
(Efficiency)	774	826	87	705	241	330	270
ALTERNATIVE 3	9.8	10.1	1.3	8.6	2.7	4.9	3.1
(% Reduction)	27.7%	14.0%	42.2%	28.5%	39.4%	20.9%	20.9%
(Efficiency)	1251	544	307	1130	576	433	274
ALTERNATIVE 4	9.6	8.7	1.4	8.6	2.7	4.7	2.9
(% Reduction)	29.2%	25.4%	37.2%	28.6%	39.5%	25.3%	27.6%
(Efficiency)	1108	828	227	958	484	439	302

NOTE: Efficiency is in units of reduced deposition (in grams of sulfur per hectare per year) per reduced emissions (million tons of sulfur dioxide)

APPENDIX A

RECEPTOR: Vermont/New Hampshire

Emission Reduction Scenario	Deposition Predicted by Transfer Matrix (kg total sulfur/ha/yr)						
	AES	ASTRAP	ENAMAP	MCARLO	MEP	OME	UMACID
CURRENT	24.7	23.0	13.9	24.2	9.7	12.1	8.5
SENATE BILL	19.5	16.3	10.7	17.9	8.0	9.6	6.3
(% Reduction)	21.0%	29.1%	23.0%	26.0%	18.0%	20.9%	25.6%
(Efficiency)	519	666	319	627	174	252	217
SIKORSKI/WAXMAN	19.3	15.9	10.5	17.6	7.8	9.5	6.2
(% Reduction)	22.0%	30.6%	24.3%	27.3%	19.3%	21.8%	26.9%
(Efficiency)	523	676	324	636	181	254	219
D'AMOURS	18.5	15.1	9.9	16.7	7.6	9.1	5.8
(% Reduction)	25.0%	34.4%	28.7%	31.1%	22.2%	24.5%	31.1%
(Efficiency)	510	652	328	620	178	244	217
STAFFORD	18.5	14.9	10.0	16.6	7.6	9.1	5.9
(% Reduction)	25.3%	35.0%	27.7%	31.3%	21.7%	25.1%	30.9%
(Efficiency)	520	668	320	629	175	253	218
UNIFORM	18.5	15.2	9.9	16.7	7.2	9.3	5.9
(% Reduction)	25.1%	33.8%	28.7%	30.9%	25.5%	23.1%	30.5%
(Efficiency)	621	776	399	748	248	280	259
ELIM. POWER PLANT SO_2	14.9	11.2	8.2	12.8	6.3	7.4	4.5
(% Reduction)	39.6%	51.1%	40.6%	47.2%	35.6%	38.7%	47.3%
(Efficiency)	581	696	334	677	205	278	238
ALTERNATIVE 1	19.7	15.6	9.9	17.9	7.8	9.7	6.6
(% Reduction)	20.2%	32.1%	28.5%	26.0%	20.0%	20.0%	22.6%
(Efficiency)	1017	1499	804	1275	396	492	390
ALTERNATIVE 2	19.8	15.6	10.3	17.9	7.9	9.8	6.5
(% Reduction)	19.9%	32.0%	25.8%	26.0%	19.1%	20.4%	22.7%
(Efficiency)	820	1224	596	1045	310	410	321
ALTERNATIVE 3	18.7	14.9	10.2	17.0	7.1	9.0	6.4
(% Reduction)	24.2%	35.2%	26.4%	29.7%	26.5%	25.7%	24.9%
(Efficiency)	1983	2681	1217	2381	855	1032	700
ALTERNATIVE 4	15.0	14.4	8.0	15.4	5.6	7.3	5.9
(% Reduction)	39.2%	37.2%	42.1%	36.3%	42.2%	39.6%	30.7%
(Efficiency)	2700	2378	1630	2450	1144	1335	726

NOTE: Efficiency is in units of reduced deposition (in grams of sulfur per hectare per year) per reduced emissions (million tons of sulfur dioxide)

APPENDIX A

RECEPTOR: Adirondacks

Emission Reduction Scenario	Deposition Predicted by Transfer Matrix (kg total sulfur/ha/yr)						
	AES	ASTRAP	ENAMAP	MCARLO	MEP	OME	UMACID
CURRENT	27.6	29.9	18.8	34.3	9.7	13.3	12.6
SENATE BILL	19.9	20.3	13.7	23.6	7.3	9.6	8.5
(% Reduction)	27.9%	32.1%	27.7%	31.3%	24.1%	27.8%	32.8%
(Efficiency)	769	961	511	1071	233	370	413
SIKORSKI/WAXMAN	19.6	19.9	13.4	23.1	7.2	9.5	8.3
(% Reduction)	29.1%	33.6%	28.6%	32.6%	25.5%	29.0%	34.1%
(Efficiency)	771	969	516	1076	237	372	414
D'AMOURS	18.5	18.6	12.6	21.6	6.9	9.0	7.8
(% Reduction)	32.9%	37.8%	32.7%	36.9%	29.1%	32.3%	38.4%
(Efficiency)	748	935	507	1043	232	355	399
STAFFORD	18.3	18.3	12.6	21.4	6.8	8.9	7.6
(% Reduction)	33.7%	38.7%	32.9%	37.7%	29.2%	33.5%	39.5%
(Efficiency)	771	964	512	1073	234	371	414
UNIFORM	18.6	18.4	12.5	22.0	6.6	9.2	8.2
(% Reduction)	32.4%	38.5%	33.4%	35.8%	31.7%	30.6%	35.1%
(Efficiency)	895	1154	627	1229	307	409	443
ELIM. POWER PLANT SO_2	13.5	12.9	9.5	15.2	5.2	6.4	5.4
(% Reduction)	51.2%	57.9%	49.5%	55.5%	46.1%	52.3%	57.3%
(Efficiency)	837	1012	551	1130	265	414	428
ALTERNATIVE 1	20.0	18.2	12.5	23.6	6.8	9.5	8.8
(% Reduction)	27.5%	39.1%	33.4%	31.3%	29.3%	28.8%	29.9%
(Efficiency)	1545	2380	1276	2176	577	783	766
ALTERNATIVE 2	20.1	18.3	12.9	23.6	7.0	9.5	8.8
(% Reduction)	27.2%	38.8%	31.1%	31.3%	27.9%	28.9%	30.2%
(Efficiency)	1251	1937	973	1785	450	643	634
ALTERNATIVE 3	18.6	17.5	11.4	23.6	6.0	8.9	9.1
(% Reduction)	32.5%	41.6%	39.2%	31.3%	37.8%	33.4%	27.5%
(Efficiency)	2975	4131	2443	3550	1213	1477	1148
ALTERNATIVE 4	18.7	19.1	11.4	23.6	6.0	8.8	9.2
(% Reduction)	32.1%	36.1%	39.2%	31.3%	38.3%	34.2%	27.2%
(Efficiency)	2469	3016	2052	2992	1033	1271	955

NOTE: Efficiency is in units of reduced deposition (in grams of sulfur per hectare per year) per reduced emissions (million tons of sulfur dioxide)

APPENDIX B
FIFTEEN MAJOR CONTRIBUTORS TO TOTAL SULFUR DEPOSITION IN THE ADIRONDACKS

AES		ASTRAP		ENAMAP		MCARLO		MEP		OME		UMACID	
Ontario	16.4%	N.Y.	31.1%	N.Y.	25.1%	N.Y.	18.4%	N.Y.	22.1%	Ontario	20.5%	Ohio	16.8%
N.Y.	15.7%	Pa.	16.6%	Pa.	17.1%	Pa.	16.7%	Pa.	15.5%	N.Y.	12.5%	Ontario	15.1%
Pa.	15.1%	Ohio	10.8%	Ontario	13.5%	Ohio	12.6%	Ontario	14.8%	Ohio	11.4%	Pa.	13.3%
Ohio	11.7%	Ontario	10.2%	Ohio	11.8%	Ontario	11.5%	Quebec	13.2%	Pa.	10.1%	N.Y.	11.9%
Quebec	9.8%	Ind.	5.7%	Quebec	11.2%	Quebec	6.9%	Ohio	9.0%	Mich.	6.3%	Ind.	6.0%
W. Va.	5.3%	Mich.	3.4%	W. Va.	6.1%	W. Va.	5.5%	W. Va.	3.8%	Ind.	5.1%	W. Va.	4.7%
Mich.	3.7%	Ill.	3.0%	Mich.	2.6%	Ind.	4.7%	N.J.	2.9%	Ill.	3.4%	Mich.	4.4%
Ind.	3.5%	Quebec	2.7%	N.J.	1.9%	Mich.	3.8%	Mass.	2.9%	W. Va.	3.2%	Mass.	3.6%
Mass.	2.3%	W. Va.	2.7	Ind.	1.8%	Ill.	2.4%	Md.	2.7%	Wis.	2.2%	Ill.	3.0%
N.J.	2.0%	Ky.	2.2%	Md.	1.7%	Mass.	2.2%	Mich.	2.6%	Mo.	1.8%	Ky.	2.4%
Ill.	1.9%	Mo.	2.0%	Mass.	1.5%	Ky.	2.0%	Ind.	1.9%	Mass.	1.8%	Tenn.	1.5%
Md.	1.9%	Mass.	1.1%	Ky.	1.2%	N.J.	1.5%	N.Y.	1.1%	Ky.	1.7%	Mo.	1.5%
Ky.	1.6%	Tenn.	1.1%	N.H.	0.8%	N.C.	1.4%	Conn.	0.9%	Iowa	1.2%	N.J.	1.2%
N.C.	1.3%	Wis.	1.0%	Tenn.	0.7%	Md.	1.4%	Ill.	0.9%	Tenn.	1.0%	Wis.	1.1%
Tenn.	1.0%	Md.	1.0%	Del.	0.5%	Mo.	1.1%	Ky.	0.7%	Minn.	0.9%	Conn.	1.1%

7
The Design of Cost-Effective Strategies to Control Acidic Deposition

David G. Streets

INTRODUCTION

Since 1981 the U.S. Congress has been considering whether to legislate a control program to alleviate the perceived effects of acidic deposition. During the four-year span of the 97th and 98th Congresses, 37 bills were proposed and debated in congressional committees without a single one reaching a vote in the House or Senate. The proposals represented some of the most significant and costly pieces of environmental legislation ever contemplated which, if implemented, would have had important ramifications for energy and environmental policy over the next several decades. However, Congress was unable to chart a course of action during this long period of deliberation.

THE CONGRESSIONAL DILEMMA

The technical problems facing Congress in regard to acidic deposition are rooted in scientific uncertainty and the absence of culpability. Although scientific research in the last decade has given important clues to the nature of the acid rain phenomenon, there is still insufficient evidence to identify the causes of the damage that has been observed, the mechanisms of damage, the extent of damage, or the risks to sensitive ecosystems now and in the future. This scientific uncertainty arises primarily because observed effects are the result of low levels of exposure over long time-periods in the presence of confounding influences. Long-term monitoring data do not exist, experimental findings are of limited applicability, and modeling techniques are rudimentary. Congress has been told there is a problem but has not been given the tools to solve it.

The political problems can be gleaned by reference to Figure 7.1, which summarizes geographical

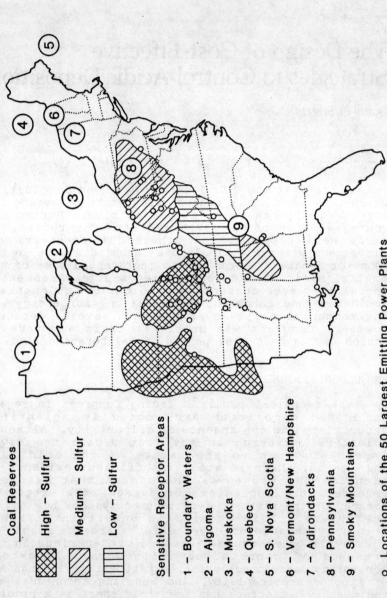

FIGURE 7.1
Geographical dimensions of the acid deposition issues in the eastern United States

Coal Reserves

▨ High – Sulfur
▧ Medium – Sulfur
▤ Low – Sulfur

Sensitive Receptor Areas

1 – Boundary Waters
2 – Algoma
3 – Muskoka
4 – Quebec
5 – S. Nova Scotia
6 – Vermont/New Hampshire
7 – Adirondacks
8 – Pennsylvania
9 – Smoky Mountains

o Locations of the 50 Largest Emitting Power Plants

information on the major sources of sulfur dioxide emissions, locations of regions particularly sensitive to acidic deposition, and major coal reserves in the eastern United States. The belt of industrialized states surrounding the Ohio River Basin contains most of the major emitting sources, which burn predominantly high-sulfur coal from local midwestern or northern Appalachian coal fields. The sensitive receptor areas are mostly in the northeastern United States and southeastern Canada. A major control program would produce benefits in these receptor regions, while costs would be incurred preferentially in the Midwest. Coal-burning plants would tend to switch to lower-sulfur coals from the West and central Appalachia. Regional conflict is inevitable; and the absence of definitive scientific findings provides opportunity for regional interest groups to find evidence, however anecdotal or conjectural, to support their positions. The scientific uncertainty fuels political conflict.

For these reasons, in 1980 Congress established an Acid Precipitation Task Force (Energy Security Act, 1980) and a ten-year research program "to identify the causes and effects of acid precipitation." A second and equal objective of the program was to "identify actions to limit or ameliorate the harmful effects of acid precipitation." This leads to the second major problem facing Congress: lack of culpability.

It is known that emissions of sulfur dioxide and nitrogen oxides from manmade stationary sources are a major contributor to the acidity of rain in the eastern United States, but the roles of other pollutants (primarily oxidants) and other emission sources (transportation, natural sources, etc.) are uncertain. The relationship between emissions, deposition, and damage is also poorly understood. The most troublesome aspect of all this to Congress is that, unlike the situation with local air pollution and most other environmental problems, it is not possible to identify culpable sources of emissions with any degree of confidence. Source-signature experiments are infeasible at present, and no theoretical model is sufficiently resolute to be able to determine an accurate relationship between emissions from individual sources and deposition at specific receptor sites. What remains is a regional-scale problem, in which each of many different sources contributes in an unknown way to acidic deposition to a particular receptor site. This uncertainty causes unprecedented difficulties for Congress to design a defensible control program within the framework of the Clean Air Act--a piece of legislation aimed at identifying and controlling specific individual sources responsible for local air pollution problems.

TABLE 7.1
Summary of 1980 emissions of the two major acid rain precursors by source category and region

Source Category	SO_2 Emissions (10^3 tons/yr)			NO_x Emissions (10^3 tons/yr)		
	East[a]	West	Total	East[a]	West	Total
Electric utilities	16,070	1,260	17,320	5,870	2,250	8,120
Nonutility combustion	3,760	820	4,580	2,420	2,770	5,190
Nonferrous smelters	110	1,110	1,220	0	0	0
Transportation	550	320	870	6,100	3,000	9,100
Other sources	1,820	1,290	3,110	670	590	1,270
Total	22,310	4,800	27,100	15,060	8,610	23,670

Totals may not add due to independent rounding.

[a]Eastern 31-state region and District of Columbia.

Source: Development of the NAPAP Emission Inventory for the 1980 Base Year, Engineering-Science Report for U.S. EPA (June 1984).

The processes of scientific investigation and public policymaking are unfinished; many key pieces of information are missing. However, this may not be enough to hinder the legislative process. Congress may decide that the risks of delaying action outweigh the value of complete information. For this reason it is important to examine the control options presented to Congress using currently available, admittedly imperfect, analytical techniques. Such techniques, while not able to project precisely the effects of any given control program, should be useful for discerning the relative merits of different programs.

Any control program that seeks to reduce emissions sufficient to relieve the acid deposition problem will be costly. It will undoubtedly have repercussions for the coal industry in the Midwest and the Appalachian region. It will cause the utility industry to reevaluate its thinking on system planning, fuel choice, and technology mix. Reverberations will be felt in the manufacturing industries and in the pocket of the American homeowner. However, if Congress resolves that such impacts must be borne in order to protect the more fragile ecosystems and valuable materials in the environment, the environmental benefits must be gained at minimum cost and with minimum disruption to society. It is the purpose of this contribution to highlight cost-effective or efficient ways to achieve the goal of reduced damage form acidic deposition. It will be shown that there are several options that offer considerably improved efficiency compared to those currently being considered by Congress.

Methods and Models

Faced with uncertainty over which sources are responsible for observed damage, Congress turned to a different but more visible target: sources responsible for the greatest emissions. Table 7.1 shows the magnitude of sulfur dioxide (SO_2) and nitrogen oxide (NO_x) emissions in the United States by source category and by region. Both species contribute to acidity in precipitation. Electric utilities in the East are clearly responsible for the bulk of SO_2 emissions (60%), whereas NO_x emissions are more evenly distributed among source categories and regions. Partly for this reason and partly because they are considered easier to control than other types of source, coal-fired electric utility power plants in the East have been the focus of control proposals. Control of SO_2 emissions rather than NO_x emissions is stressed because of the greater potential reduction and the wider availability of demonstrated control technologies.

Fortunately, the knowledge of emissions, control opportunities, and control costs is also greatest in this area. Several reputable models have been developed in recent years to project the control costs and emission reductions associated with hypothetical control programs for electric utilities. Notable among these are the Coal and Electric Utilities Model (CEUM) [ICF, Inc., 1983], a proprietary model developed by ICF Incorporated from the National Coal Model (NCM). The U.S. Department of Energy's (DOE's) Energy Information Administration has used a system based on the NCM and the National Utility Financial Statement (NUFS) model (Energy Information Agency, 1983). Teknekron, Inc., developed the Utility Simulation Model (USM) [Teknekron Research, Inc., 1981], which was used jointly with CEUM in several studies of emissions reduction strategies in the late seventies. Work is now in progress on an unhanced version of USM called the Advanced Utility Simulation Model (AUSM) [Universities Research Group on Energy, 1983]. The AUSM should advance state-of-the-art utility system modeling, and some preliminary results have already been presented (Stukel and Bullard, 1984; and Bullard and Hottman, 1984). When fully developed, the AUSM will form the primary emissions and cost modeling tool of the National Acid Precipitation Assessment Program (NAPAP). Argonne National Laboratory and E. H. Pechan and Associates have modified and applied the AIRCOST model (E. H. Pechan & Associates, Inc., 1983), the pollution control cost module from USM. While sacrificing some of the detail of utility system operation, the AIRCOST model has proved useful in policy evaluation of potential acid deposition control strategies. Representatives of the electric utility industry (Temple, Barker & Sloane, Inc., 1984; and Damon, et. al., 1984) and congressional support groups (Parker and Kumins, 1983; and OTA, 1983) have used simpler methods, based on experience with particular components of the problem rather than on all-encompassing models.

The discussion and results that follow for electric utilities are based on work performed at Argonne National Laboratory. However, the methods used and the results obtained by the different groups tend to be similar; there is no great methodological debate. Where similar studies to the ones described here have been performed by other groups, they are referenced. Limited analysis of nonutility-sector emission reductions has been performed by several groups and is noted accordingly. Almost all analysis has focused on SO_2 emission reductions rather than NO_x.

What general principles are involved in evaluating strategies to reduce SO_2 emissions from electric

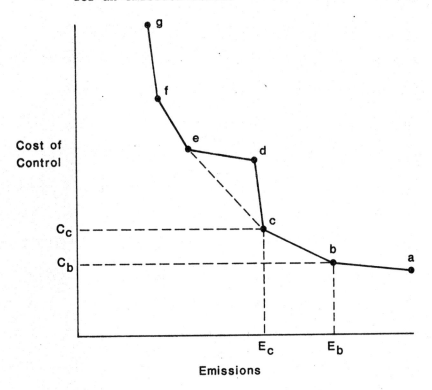

FIGURE 7.2
Hypothetical control-cost function for an emission source

utility power plants? The analytical process is driven by two questions: what overall magnitude of emission reduction is desired, and what magnitude of reduction is desired at each plant? Two corollary questions motivate the design of an analytical methodology: what is the least-cost way of achieving a desired level of reduction at each plant? And what is the least-cost way of achieving a desired level of total reduction among all plants? As the next section shows, current proposals before Congress have addressed the first two questions, but usually without recourse to model predictions; as the rest of this chapter shows, answering the second two questions enables efficiency improvements to be made.

All of the modeling systems fundamentally operate by determining the least-cost option for a particular generating unit to achieve a specified reduction in SO_2 emissions. Options considered are fuel cleaning, fuel blending, fuel switching, and installation of control technology (typically wet and dry flue-gas desulfurization). The more sophisticated models also take into account dispatching of units and selective retirement of units. By considering a range of emission reduction levels, a complete least-cost reduction for each unit does not exist, and if other constraints are taken into account, such curves can be used to determine least-cost solutions among a group of plants.

Finding a cost-effective emission control solution among a group of plants is not simply a matter of deciding which sources to control and which not to control. Instead, the most cost-effective level of control for each source must be determined. Emissions should be reduced at each source until further reduction is not cost-effective. In practice, an overall emissions target is specified and stepwise reductions in emissions are made among the units until the target level of emissions is achieved. In this way, any emission goal can be met in the most cost-effective manner.

To find a least-cost solution for a group of sources, curves of control costs vs. emissions must be constructed for each individual source. A typical cost function for a source is shown in Figure 7.2. The curve is piecewise linear, that is, it consists of connected line segments. The slopes of these line segments are the incremental costs (in dollars per ton of SO_2 removed).

For example, the incremental cost IC of reducing emissions from E_b to E_c is the slope of the line segment joining points b and c:

$$IC_{VC} = \frac{C_c - C_b}{E_c - E_b},$$ where C is the total cost of control.

The least-cost solution will never correspond to a point such as d, which is inside the dotted line connecting points c and e, because emissions can be reduced all the way from point c to e at a lower incremental cost (slope) than from point c to d. That is, whenever it would be cost-effective to reduce emissions to d, it would also be cost-effective to continue to reduce emissions to e or beyond.

Points like d arise frequently. For example, emissions might be reduced to point c by physical coal cleaning, which is relatively cheap. Further reductions, however, may require a new technology. Perhaps flue-gas desulfurization (FGD) equipment must be installed to reach point d. Support point d represents 70% SO_2 removal with FGD. Once FGD equipment is installed, the incremental cost of further emission reduction is relatively low. Point e, therefore, might represent 80% SO_2 removal with FGD. This hypothetical example suggests that 80% SO_2 removal with FGD would be more cost-effective than 70% SO_2 removal. Argonne has developed a procedure by which a postprocessor model, MINCOST, takes the output of several runs of the unit-level control model, AIRCOST, and determines the least-cost control option among many units. MINCOST is based on five steps:

1. Construct the control cost functions for each source on the basis of AIRCOST output, by incrementally varying the emissions goal for each source. This step is equivalent to determining points a-g in Figure 7.2.

2. Compute for each source the incremental costs of moving between each adjacent pair of points. The incremental cost is the slope of the line segment connecting any two points.

3. If the incremental costs do not uniformly increase with reductions in emissions, make appropriate corrections (e.g., in Figure 7.2, skip points like d and compute incremental costs between points like c and e). This step is termed "construction of the convex hull."

4 Arrange the control options among all the units in order of increasing incremental costs. Apply controls up to the point where

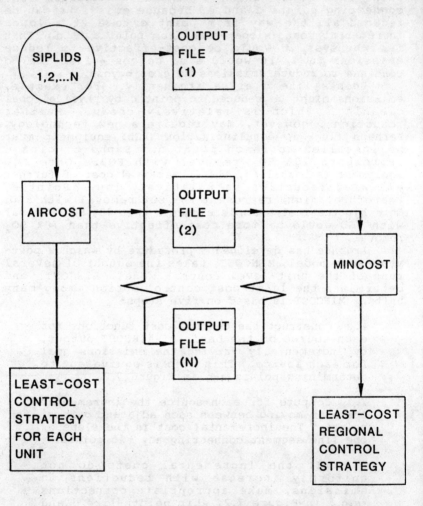

FIGURE 7.3
Methodology for determining the least-cost strategy for a group of sources

the overall emissions target is met. The incremental cost of further emission reduction is called the threshold incremental cost IC. Sum the unit-level emissions and costs.

5. Vary IC or the overall emissions goal to yield least-cost Control curves for all sources to achieve the stated objectives of a control program.

Figure 7.3 summarizes the computational procedure. The SIPLIDS are the successively more stringent emission limitations placed on individual units (lids, or ceilings, on emissions allowed under state implementation plans) to simulate points a-g. This general methodology is used by all modeling systems and specifically by the one that generated the results described herein.

Beyond the realm of SO_2 reductions from utility power plants, analytical capabilities are more primitive. ICF Incorporated has developed an Industrial Boiler Model (IBM); and EEA, Inc., is developing the Industrial Combustion Emissions (ICE) model for the National Acid Precipitation Assessment Program. Knowledge of control opportunities for nonferrous smelters is available on a source-specific basis because they are relatively few in number. Apart from this, Argonne National Laboratory and E. H. Pechan and Associates (Schwengels, et. al., 1984) have developed relatively unsophisticated models that attempt to capture SO_2 reduction opportunities in all major emitting sectors, but there is room for more work in this area. Preliminary investigations suggest that the control opportunities outside the electric utility sector are limited and relatively expensive; moreover, Congress is unlikely to target these sectors, because sources are small, dispersed, and vulnerable to additional capital expenditure requirements. In addition, opportunities for control of nitrogen oxide emissions from all sectors should be examined. Analytically, this is difficulty because of the site-specific nature of NO_x emissions and the uncertain cost and performance of current and developing control technologies. Here also is an area for fruitful investigation in the future.

ECONOMIC EFFICIENCY

To address the acidic deposition control issue, 14 bills were introduced into the 97th Congress, and 23 bills were introduced into the 98th Congress (Streets, et. al., January 1983). Most of these bills call for reductions in annual SO_2 emissions of 8-12 million tons. As Table 7.1 shows, this level of emission reduction represents 50-75% of 1980 SO_2 emissions from electric utility power plants in the eastern United States, 35-55% of total SO_2 emissions in the eastern United States, and 30-45% of total SO_2 emissions nationwide.

This choice of the amount of emission reduction is derived largely from a statement by the National Research Council that "it is desirable to have precipitation with pH values no lower than 4.6 to 4.7 throughout such [sensitive freshwater] areas.... In the most seriously affected areas...this would mean a reduction of 50% in deposited hydrogen ions" (NRC, 1981). It has been assumed that 50% reduction in emissions of acid-causing pollutants (in this case SO_2) would reduce hydrogen ion deposition by approximately 50%. This statement may be true only for certain geographical and temporal scales and for certain climatological conditions. Nevertheless, it forms the basis of current legislative proposals.

Most bills take a broad approach to reducing SO_2 emissions, defining an acceptable emission rate (usually 1.2 lb $SO_2/10^6$ Btu) for a source (usually a power plant) and allocating emission reductions to states in proportion to current emissions in excess of the acceptable emission rate. States are then given the responsibility for determining which sources within the state should control emissions further. Even if each state allocates reductions to sources within that state in a cost-effective manner, allocations to states are not necessarily efficient.

The allocation procedures do not consider the geographical locations of sources that would reduce emissions. Thus, if the control programs have no preference for which states achieve reductions, it seems reasonable to consider the different costs of reducing SO_2 emissions in different states. The required emission reduction could be achieved at lower cost by optimizing over the entire region, that is, by reducing emissions more in states where control costs are lower. This strategy is equivalent to an interstate emissions trading option. As discussed earlier, such an option can be simulated by merging unit-level cost data and recomputing the minimum costs to achieve the same level of emission reduction as the bills require. Comparing the costs of the bills with the least-cost alternative yields a measure of the economic efficiency of the proposed programs. The efficiency varies with the particular proposal, so a brief examination of several of the more important bills is necessary.

One of the most important bills of the 98th Congress was H.R. 3400, introduced by Reps. Gerry Sikorski (D-MN), Henry A. Waxman (D-CA), and Judd Gregg (R-NH). Commonly known as the Sikorski-Waxman bill, H.R. 3400 would require a nationwide reduction in SO_2 emissions of 10 million tons/yr, to be achieved partially through retrofitting continuous emission control devices on the 50 largest power-plant emitters (the locations of which are shown in Figure 7.1). Approximately four million tons reduction in NO_x emissions would be achieved by tightening New Source Performance Standards for power plants and strengthening truck emission limits. The Sikorski-Waxman bill also introduced the concept of an acid deposition control fund. In essence, utilities would pay into the fund at the rate of 1 mill/kWh of electricity generated, and payments would be disbursed from the fund to pay for 90% of the capital costs of control systems. An acid deposition control fund may be necessary for cost-effective control strategies, which tend to place a large burden on relatively few states where control costs are low.

Two important bills followed H.R. 3400. Norman D'Amours (D-NH) introduced H.R. 4404 to extend and amend several features of the Sikorski-Waxman proposal. An additional reduction of two million tons of SO_2 would be required, to be obtained from industrial fuel combustion and industrial process sources. The fee to be paid into the control fund would be raised to 1.5 mill/kWh, indexed to inflation.

Representatives Morris K. Udall (D-AZ) and Dick Cheney (R-WY) introduced H.R. 5370 as a "freedom-of-choice" alternative to H.R. 3400. The Udall-Cheney bill took a two-phased approach to SO_2 emission reduction: 5.5 million tons/yr by 1991 and 11 million tons/yr by 1996. The bill rejects both the forced scrubbing and control fund concepts of the Sikorski-Waxman bill. State allocations would be determined by agreement of the governors of states in the region. Failing agreement, an allocation based on excess emissions above 1.2 lb $SO_2/10^6$ Btu would apply. This bill can be analyzed on the basis of either the allocation formula (intrastate emissions trading) or the assumption that the governors agree on a cost-effective allocation (interstate emissions trading).

At the beginning of the first session of the 98th Congress, Sen. Robert T. Stafford (R-VT) introduced S.768, which was a reintroduction of the bill approved by the Senate Committee on Environment and Public Works in the 97th Congress. This bill required a reduction in annual SO_2 emissions of eight million tons by 1993, with no restrictions on choice of control technology. State shares of the total reduction were based on

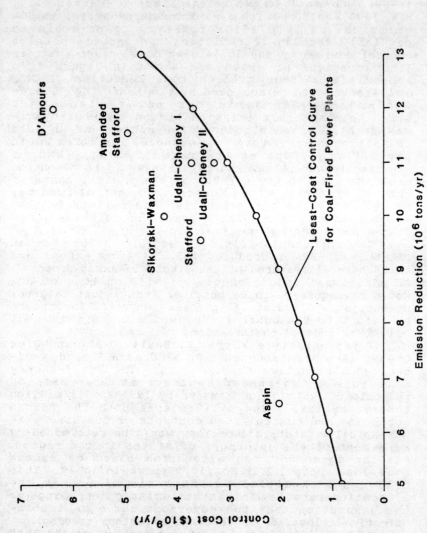

FIGURE 7.4 Comparison of the costs of alternative control proposals with the least-cost curve

emissions in excess of 1.5 lb $SO_2/10^6$ Btu. The Stafford bill was subsequently amended in committee to require a reduction of 10 million tons by 1994, with state shares based on emissions in excess of 1.2 lb $SO_2/10^6$ Btu. After inclusion of additional reductions to offset new growth, the reductions required by these two versions of the Stafford bill are 9.6 and 11.6 million tons/yr, respectively.

These bills are the most important fom the 98th Congress. Argonne National Laboratory has analyzed the costs and state-level emission reductions of these bills using the AIRCOST model. Several of these bills have also been analyzed by other groups. All these findings are compiled in Table 7.2. Figure 7.4 compares the Argonne costs for these bills with the least-cost control curve for coal-fired power plants determined with the MINCOST model. This curve indicates the minimum cost to achieve a given level of SO_2 reduction by selecting the lowest-cost reduction opportunities among all units in the region. Also shown in Figure 7.4 is the Aspin bill discussed later in this chapter.

The vertical separation of each point corresponding to a particular bill from the least-cost curve is a measure of the cost inefficiency of the proposal. The two versions of the Udall-Cheney bill are the most efficient because they allow utilities flexibility to choose among several alternative methods to achieve the mandated reduction in emissions. The interstate trading version of the Udall-Cheney bill is the most efficient because it does not constrain the states to an inefficient allocation formula. It would actually lie on the curve, if not for small inefficiencies caused by the two-phased approach to control, which locks certain utilities out of efficient strategies in Phase II because of options chosen in Phase I. This inefficiency is estimated to cost $0.22 billion/yr. The intrastate trading version of the Udall-Cheney bill is less efficient because of the state allocation formula. It is estimated to cost $0.68 billion/yr more than the least-cost method of achieving that level of emission reduction.

The two versions of the Stafford bill are less efficient than the Udall-Cheney bill. Although the Stafford bill permits several options for controls, the same as in the Udall-Cheney bill, it has the added requirement of growth offsets. Offsetting growth in the states in which it occurs is an inefficient way to reduce total emissions compared to achieving an equivalent reduction in emissions in states where the marginal cost of reductions beyond the amounts specified in the bill is relatively low. Thus the original Stafford bill costs about $1.25 billion/yr

TABLE 7.2
Comparison of emission reductions and costs (1980 dollars) to comply with alternative control proposals

Control Proposal	SO$_2$ Emission Reduction (10^6 tons/yr)	Control Cost ($10^9/yr)	Least-Cost Strategy ($10^9/yr)	Cost Inefficiency Measure Difference ($10^9/yr)	Percent of Least-Cost
Sikorski-Waxman	10.00	4.22[a]	2.48	1.74	170
D'Amours	12.00	6.37	3.68	2.69	173
Udall-Cheney I[b]	11.00	3.72	3.04	0.68	122
Udall-Cheney II[c]	11.00	3.26	3.04	0.22	107
Aspin	6.52	2.63	1.20	1.43	219
Stafford	9.57	3.53[d]	2.28	1.25	155
Stafford, as amended	11.57	4.92	3.38	1.54	146

[a] Ref. 12 estimates control cost as $3.7–4.4 × 10^9/yr; Ref. 9 estimates control cost as $5.1 × 10^9/yr.

[b] Assuming intrastate trading according to state allocation formula.

[c] Assuming interstate trading agreement by governors of states in the region.

[d] Ref. 2 estimates control cost as $3.1 × 10^9/yr; Ref. 3 estimates control cost as $3.6 × 10^9/yr.

more than the least-cost method of achieving a reduction of 9.6 million tons/yr; the amended Stafford bill costs about $1.54 billion/yr more than the least-cost method of achieving a reduction of 11.6 million tons/yr.

The Sikorski-Waxman bill is inefficient in its achievement of a 10 million ton reduction in emissions, costing $1.74 billion/yr more than the least-cost option. This inefficiency clearly arises because of the restricted, forced scrubbing requirement imposed on the 50 largest emitters. Cheaper options for these plants that might involve fuel switching are not permitted.

The D'Amours bill is the most inefficient bill of all. It combines the inefficiencies of the Sikorski-Waxman bill with a costly requirement to reduce industrial combustion emissions and industrial process emissions. It is more costly to achieve reductions in the industrial sector than to achieve equivalent reductions from utility plants in those states where the marginal cost of achieving reductions beyond those required under the Sikorski-Waxman formula is relatively low. The D'Amours bill is estimated to cost $2.69 billion/yr more than the least-cost strategy to achieve a 12 million ton reduction. Clearly, if states under the D'Amours bill were permitted to trade cheaper utility emission reductions for more-expensive industrial emission reductions, the costs would decrease. Table 2 shows the measures of inefficiency expressed as absolute and percentage differences relative to the least-cost option.

ICF Incorporated has also analyzed cost-effective strategies for SO_2 emissions (ICF, Inc., 1984). Their major findings are threefold: (1) costs of reducing SO_2 emissions increase sharply as the level of reduction increases, nearly doubling for every additional two million tons of reduction beyond six million tons; (2) fuel switching is generally the most cost-effective means of reducing SO_2 emissions, with corresponding impacts on regional coal production; and (3) retrofitting FGD systems is generally not cost-effective until the total reduction requirement exceeds eight million tons.

Figure 7.4 compares the economic efficiencies of different bills. It shows that the congressional proposals are generally far from efficient solutions to the problem of reducing SO_2 emissions. However, it does not capture the full range of attributes. For example, the greater the efficiency of a bill, the greater the fuel switching that occurs, and the greater the disruption of existing patterns of coal supply. Representatives Sikorski and Waxman were prepared to sacrifice some economic efficiency to protect coal markets. Until a method is devised to weigh a third parameter, regional coal use, against the two parameters of cost and emission reduction, it will not be possible to select the "best" emission reduction program from a slate of alternatives. All of the preceding discussion assumes, as the bills do, that location of sources reducing emissions is immaterial.

GEOGRAPHICAL EFFICIENCY

Most programs proposed to control acidic deposition fail to consider the relative contributions of emissions from different states to acidic deposition at sensitive receptor sites. This may be because atmospheric transport models are not presently considered verifiable to the degree necessary to support quantitative conclusions about the effects of emission control strategies. However, these models can be useful in discerning the relative merits of different strategies at broad levels of source-receptor aggregation.

The key to geographical efficiency lies in the fact that not all locations in the United States are equally sensitive to acidic deposition. Indeed, evidence suggests that low levels of deposition of nitrogen and sulfur in the Midwest may increase productivity of certain field crops through a fertilizer effect. The high alkalinity of soils in this and other regions is sufficient to buffer the deposited acidity. Thus, in a cost-benefit sense, it is possible that reduced deposition in certain locations may yield only a small positive benefit or, conceivably, a negative benefit. On the other hand, other locations are known to be sensitive to acidic deposition because of the low buffering capacity of soils. The sensitivity varies among regions and among receptor types.

Work Group 2 of the U.S.-Canada Transboundary Air Pollution Study (Ferguson and Machta, 1982) identified nine areas as particularly sensitive to acidic deposition. These areas are shown in Figure 7.1. Might it not be sensible, at least for a near-term strategy, to select sites for emission reduction so as to maximize the reduction in acidic deposition in these nine regions and to minimize the cost? Little benefit can be gained by reducing emissions from sources on the Atlantic coast, where the bulk of pollutants is deposited in the Atlantic Ocean, or from sources in the Southwest, where the pollutants are preferentially deposited on the highly-buffered Midwest farmland. By targeting the strategy on particular sensitive regions, improvements in cost-effectiveness are possible.

TABLE 7.3
Quantitative findings of the targeted
strategy analysis

Strategy	Annual Control Cost[b] ($10^9)	Reduction in Annual SO$_2$ Emissions[c] (10^6 tons)	Reduction in Total Sulfur Deposition in the Adirondacks[d] (%)
Senate Bill S.768			
As proposed	4.2	9.0	23 (21-26)
Cost-optimized	1.9	9.0	26 (26-31)
Deposition-optimized	0.9 (0.4-1.1)	5.0 (1.9-5.5)	23 (21-26)
Mandatory FGD	4.6	9.0	23
Mandatory FGD (cost-optimized)	2.3	9.0	27
Senate Bill S.769			
As proposed	12.9	14.9	40 (38-46)
Cost-optimized	6.0	15.0	46 (45-57)
Deposition-optimized	3.0 (1.3-3.2)	9.8 (4.2-10.3)	40 (39-46)
Mandatory FGD	13.6	14.9	40
Mandatory FGD (cost-optimized)	7.0	15.0	47
20-kg Deposition Goal			
Including Muskoka	11.9	16.2	56
Excluding Muskoka	6.0	13.8	46
30-kg Deposition Goal	2.3	9.1	31

[a] Numbers in parentheses represent the range of results given by six atmospheric transport models. The precision of deposition change estimates is unknown, but the qualitative differences between strategies are believed to be reliable.

[b] In real, levelized 1980 dollars. Real-world costs would be expected to be greater than these theoretical least costs.

[c] In English short tons.

[d] Modeled using ASTRAP with 1980 meteorological data.

This concept is not new. Trisko (1983) recommended "a phased program of source controls, starting with nearby...sources, and expanding later to more distant...sources as necessary." Fay et al. (1983) suggested that "considerably lower costs could be realized in an emission control scheme that is the more stringent the nearer the sources are to be environmentally sensitive areas." Argonne National Laboratory has combined emission control cost models with atmospheric transport models, in an optimization framework, to examine this possibility analytically (Streets, et al., 1984a). Because of the uncertainty associated with current models of atmospheric transport and deposition, the study was repeated with transfer matrices derived from six different state-of-the-art models. One specific model, the Advanced Statistical Trajectory Regional Air Pollution (ASTRAP) model (Shannon, 1981), was chosen for sensitivity studies, because it shows good agreement with monitored wet deposition and was available to the authors.

To demonstrate the potential economies achievable by targeting emissions control, several versions of Senate bills S.768 and S.769 were evaluated in terms of annual control costs, SO_2 emission reductions, and sulfur deposition rates in the nine sensitive regions. The eight-million-ton-reduction version of S.768, as described in the preceding section, was studied. The bill S.769 is similar to S.768 but requires a reduction in annual SO_2 emissions of 12 million tons. Note that the targeted strategy analysis used a different version of the AIRCOST model than the bill analysis discussed in the preceding section. The targeted strategy also considered reductions in nonutility emissions. Thus costs reported for S.768 are slightly different in the two studies.

The results, as summarized in Table 7.3, are striking. Control measures proposed in the legislation are relatively expensive because of economically inefficient allocation of emission reductions among the states. Regional cost optimization over the eastern United States, as described earlier, can reduce costs by more than 50% while achieving the same level of SO_2 emission reduction as the unoptimized bills. Cost optimization would actually reduce sulfur deposition more in the sensitive regions examined than the bills would.

After determining the deposition reductions that would be achieved by each bill, the six transfer matrices were used to calculate the most cost-efficient way to achieve the same deposition reductions. A single region was targeted in this case: the Adirondacks. The cost and deposition models were used to determine the least-cost distribution of emission

193

FIGURE 7.5
Cost-effectiveness of alternative
strategies for reducing sulfur deposition

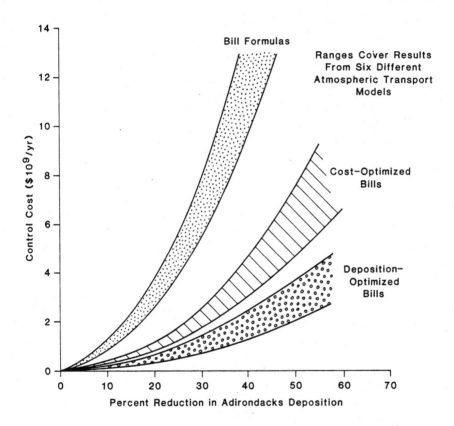

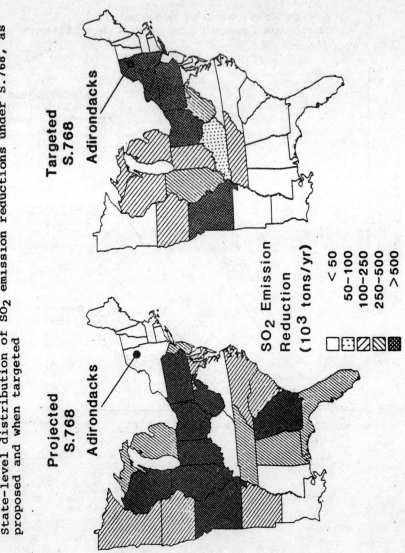

FIGURE 7.6
State-level distribution of SO_2 emission reductions under S.768, as proposed and when targeted

reductions that would reduce sulfur deposition in the Adirondacks by the same amounts as would the bills.

At an annual cost of between $0.4 billion and $1.1 billion, the targeted strategy would reduce sulfur deposition in the Adirondacks by the same amount (21-26%) as S.768, which costs 4.2 billion/yr. The ranges correspond to the results from the six different transfer matrices. Cost savings would therefore be between 75% and 90%. The necessary reduction in SO_2 emissions under the targeted strategy would be 1.9-5.5 million tons/yr, as compared with 9.0 million tons/yr under the unoptimized S.768.

Similarly, at an annual cost of $1.3-3.2 billion, the targeted strategy would reduce deposition in the Adirondacks by the same amount (39-45%) as S.769, which costs $12.9 billion/yr. Cost savings in this case are again 75-90%. The necessary reduction in SO_2 emissions under the targeted strategy is 4.2-10.3 milliion tons/yr, as compared with 14.9 million tons/yr under the unoptimized S.769.

Generally, deposition-optimized strategies perform well in the other sensitive regions in the Northeast (Quebec, S. Nova Scotia, and Vermont-New Hampshire), and to a lesser extent Pennsylvania, when optimized for the Adirondacks. However, deposition is reduced significantly less in the Boundary Waters, Algoma, and Smoky Mountains areas under the targeted strategy. Because current deposition is relatively low in the first two areas, the higher deposition may be of concern only for the Smoky Mountains region. The importance and immediacy of this concern depend on the sensitivity of ecosystems in the Smoky Mountains to acidic deposition. The strategy targeted for the Adirondacks could readily be supplemented by a second strategy targeted for the Smoky Mountains.

Figure 7.5 brings the advantages of the targeted strategy for the Adirondacks into sharp focus. It clearly illustrates the potential cost savings of a targeted approach if the objective is to reduce sulfur deposition in the Adirondacks (and, with little loss of effectiveness, in the northeastern United States and southeastern Canada). As an illustration, the targeted strategy is clearly the least expensive way to reduce deposition in the Adirondacks by, say, 40%. In contrast, a cost-optimized strategy that permits interstate trading costs about 50% more, and an unoptimized strategy that permits only intrastate trading costs about 350% more. These curves were constructed with results from the ASTRAP transfer matrix with 1980 meteorological data. Note that the range of values, covering results obtained with six different transfer matrices, does not mask the advantages of the targeted strategy. Figure 7.6

illustrates how the reduction requirements shift to states with a combination of cheap cleanup opportunities and upwind proximity to the Adirondacks, under a targeted strategy.

The targeted strategy is sensitive to meteorological variability. Estimated contributions of emissions from source regions to wet and dry deposition in receptor regions depend on the meteorological data from two periods with markedly different weather characteristics. Meteorological variability in areas close to the receptor regions is particularly critical. Consequently, targeted strategies must be based on long-term, average meteorological conditions.

Targeted strategies can also be designed to meet multiple objectives, such as achieving desired reductions in deposition in many sensitive regions at minimum cost. Analysis showed that wet deposition of sulfate could be reduced to levels below the value recommended by some scientists, 20 kg/ha-yr, in all nine sensitive regions, at slightly less than the cost of an unoptimized S.769 (which would not achieve the 20 kg/ha-yr goal in four regions). If the Muskoka region is excluded from this requirement, because deposition there is dominated by emissions from Canadian sources and the analysis did not consider reductions in Canadian emissions, then the same goal could be achieved for half the cost of an unoptimized S.769. An alternative wet deposition goal of 30 kg/ha-yr of sulfate, suggested by other scientists, could be met in all nine sensitive regions for approximately 20% of the costs to meet S.769 or the 20-kg option. This strategy also achieves the goal in Muskoka. These results are also summarized in Table 7.3.

The analysis assumed that electric utilities were allowed to reduce emissions by switching fuels, in addition to using technological emission controls. Fuel switching, however, could disrupt established markets of coal supply and adversely affect the coal industry. To avoid these impacts, control strategies, such as the Sikorski-Waxman bill, have been proposed that specifically require the use of continuous emission reduction technology (fuel-gas desulfurization). Analysis showed that control costs would increase by 6-21%, depending upon the particular control strategy considered, if fuel switching were not a permissible compliance option. Deposition patterns would not be greatly different.

A similar study was performed by Systems Applications, Inc., and ICF Incorporated (1983) for Public Service Indiana Co. They report the following significant findings: (1) states emitting the most SO_2 do not necessarily contribute the most to wet sulfate

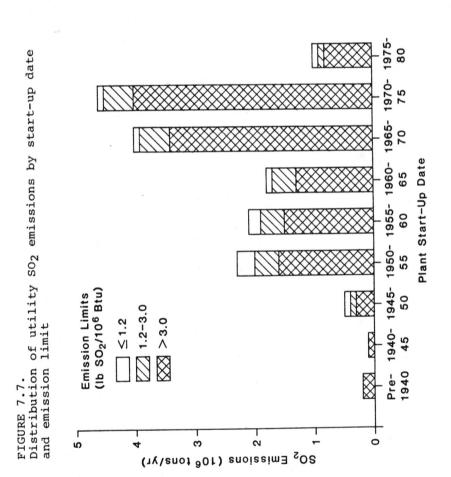

FIGURE 7.7.
Distribution of utility SO_2 emissions by start-up date and emission limit

deposition in the Adirondacks; (2) different states have widely varying costs for achieving specified emissions reductions, and cost-effectiveness in terms of wet deposition reductions per dollar decreases with increasing distance between source and target receptor area; (3) a least-cost SO_2 emission reduction approach can achieve a rate of wet sulfate deposition of 20 kg/ha in the Adirondacks at a fraction of the corresponding costs for other alternatiive emissions reduction proposals. These conclusions are entirely consistent with those of the Argonne study.

In interpreting the results of these kinds of analysis, it must be recognized that estimates from current long-range transport models are uncertain. For example, the models studied in the U.S.-Canada report on atmospheric transport varied by a factor of four in their predictions of deposition in sensitive areas, using the same emissions data and the same meteorlogy. Furthermore, the models all take a linear approach to atmospheric chemistry; in reality, the chemistry may be nonlinear. Because the models generally agree on relative control area priorities, however, they should be useful for discerning the relative merits of different control strategies. Less confidence should be placed in their absolute numerical outputs. The conclusion to be reached from this kind of study is that, if reducing sulfur (acidic) deposition in the northeastern United States and southeastern Canada is the primary objective of a control program, then a targeted approach to control has great potential cost savings.

TEMPORAL EFFICIENCY

The question of how best to limit the emissions from existing sources should not be answered without considering the ages of current sources and their expected lifetimes. These factors will strongly influence the pattern of emissions in the future and hence the effects of strategies to control such emissions. An efficient control program should aim not only to maximize emission reductions in a given future year but also to maximize cumulative emission reductions over the lifetime of the existing stock of coal-fired power plants.

A common misconception holds that acid rain is caused by older power plants that will probably retire before the end of the century and thus alleviate the problem. In fact, the greatest emissions emanate from large power plants constructed relatively recently. Figure 7.7 shows that the bulk of current SO_2 emissions (over 8 million tons) comes from plants constructed in

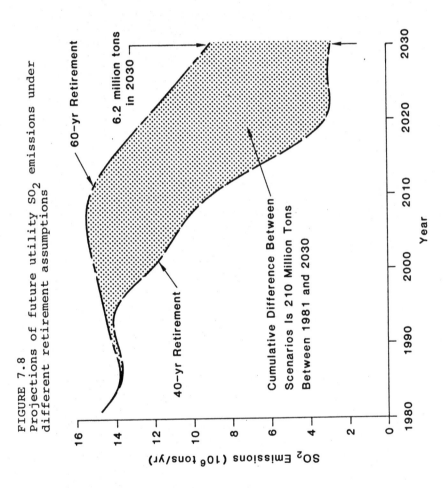

FIGURE 7.8
Projections of future utility SO_2 emissions under different retirement assumptions

the 10-year period from 1965 to 1975. Despite the passage of stringent New Source Performance Standards (NSPS) in 1971, most of these plants commenced construction before 1971, thereby gaining exemptions from NSPS.

In the absence of acid rain legislation, many of these plants will likely continue operating well into the next century. Utilities are finding it increasingly attractive to extend the lifetime of units through refurbishment and good maintenance practices, rather than to retire units and incur the high costs of new plant and emission control equipment. Lifetimes as long as 60 years are now considered possible, whereas 40-45 years was until recently considered the typical lifetime of a coal-fired power plant. Therefore, we may well see the emissions problem persistiing until 2030 or so.

Figure 7.8 shows two projections of utility SO_2 emissions out to the year 2030. One assumes that all power-plant units retire at age 40; the other assumes they all retire at age 60. The difference between them is cumulatively about 210 million tons of SO_2 over the 50-year period between 1981 and 2030. Even in the year 2030, when it might be expected that all of the heavily-emitting plants would have retired under either scenario, emissions between the two cases differ by a substantial 6.2 million tons/yr. Besides confirming that retirement age is a crucial determinant of future emissions, this result suggests that any strategy to encourage early retirement of old units would likely pay off.

If retiring capacity has to be replaced by new capacity, then it is hard to imagine utilities voluntarily participating in an early retirement program. The costs of replacement capacity (at an average capital cost of about $1200/kW) well exceed the levelized costs of refurbishment (at an average cost of about $350/kW). When differences in efficiency of operation and performance of control systems are taken into account, it is estimated that the total annual levelized cost for a new 500 MW unit would exceed by at least 25% the total annual levelized cost of extending the lives of two old 350 MW units. The preference for refurbishment over replacement is clear.

Several analysts have suggested that, because growth in electricity demand is projected to be slow in the future and because current reserve margins are high, there may be potential for retirement of some existing units without incurring expensive replacement costs. Data indicate that as much as 26 GW of additional capacity could be obtained by raising 1980 capacity utilization to a floor of 60% (Garvey et al., 1984). This increased utilization would lead to an

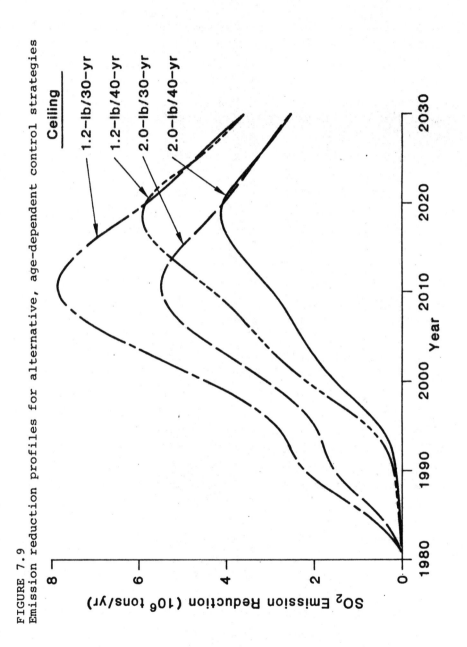

FIGURE 7.9
Emission reduction profiles for alternative, age-dependent control strategies

estimated increase in annual SO_2 emissions of 2.1 million tons, because the existing plants have higher SO_2 emission rates than the replacement capacity would have. This is of environmental concern because many bills do not make it clear that increased emissions due to increased utilization of existing plants must be offset by reductions elsewhere. In the absence of such provisions, replacement would be discouraged. However, Keelin and Oatman (1982) have refuted this notion and demonstrated that increased demand (whether through economic growth or replacement of aging capacity) would generally require construction of new capacity to ensure reliable service. Although certain individual utility companies may have excess capacity it does not appear to be a widespread phenomenon.

Of course, the concept of early retirement may not be nearly so unpalatable to utilities if the alternative is a mandatory emission reduction program. In September 1983, Representatives James T. Broyhill (R-NC) and Edward R. Madigan (R-IL) proposed phased retirement of existing plants that exceed an emission rate of 1.2 lb $SO_2/10^6$ Btu. This plan is interpreted as presenting utilities with the option of either reducing the emission rate of a unit to 1.2 lb/10^6 Btu or less when it reaches a certain age, or else retiring the unit. In some cases, retirement could be the result, although there is no way to predict utility behavior in this respect. Assuming utilities would prefer to reduce emissions to the allowable rate (because it is apparently much cheaper to do so), the time variation of emissions under alternative age-dependent control scenarios can be estimated (Garvey, et al., 1984). This process provides insight into the temporal efficiency of these scenarios.

Figure 7.8 showed the projected path of future emissions assuming a fixed 60-year lifetime of utility units. Figure 7.9 shows the time variation of emission reductions relative to this base case projection for four different control scenarios: applying a 1.2 (or 2.0) lb/10^6 Btu ceiling on plants when they reach age 30 (or 40). The scenarios show different profiles because of differences in the way they apply to the diminishing stock of existing units over time. These profiles have different implications as preferred strategies, depending on the criteria appropriate for judging efficiency. For example, a 2.0 lb/30-yr ceiling achieves a lower peak reduction than a 1.2 lb/40-yr ceiling, but achieves it ten years earlier. Which is more important: magnitude of reduction or regency? Each of these scenarios has a corresponding curve for cost of control that must also be taken into account.

One useful measure of temporal efficiency is

TABLE 7.4
Emissions reductions and increased costs to base case for our control scenarios

Scenario	Peak Year			Near-Term (1995)		
	Δ SO$_2$ (10^6 tons/yr)	Δ Costs[a] ($ 10^9/yr)	Cost-Effectiveness ($/ton)	Δ SO$_2$ (10^6 tons/yr)	Δ Costs[a] ($ 10^9/yr)	Cost-Effectiveness ($/ton)
1.2 lb ceiling/30 yr	-7.8[b]	+2.8	354	-2.9	+1.2	424
1.2 lb ceiling/40 yr	-5.8[c]	+2.0	335	-0.8	+0.4	534
2.0 lb ceiling/30 yr	-5.5[b]	+1.5	281	-2.0	+0.6	327
2.0 lb ceiling/40 yr	-4.1[c]	+1.1	267	-0.5	+0.2	402

Scenario	Long-Term (2020)			Cumulative (50 years)		
	Δ SO$_2$ (10^6 tons/yr)	Δ Costs[a] ($ 10^9/yr)	Cost-Effectiveness ($/ton)	Δ SO$_2$ (10^6 tons)	Δ Costs[a] ($ 10^9)	Cost-Effectiveness ($/ton)
1.2 lb ceiling/30 yr	-5.8	+2.0	335	-216.8	+78.4	362
1.2 lb ceiling/40 yr	-5.8	+2.0	335	-140.1	+50.5	360
2.0 lb ceiling/30 yr	-4.1	+1.1	267	-150.6	+43.4	288
2.0 lb ceiling/40 yr	-4.1	+1.1	267	-97.3	+28.2	290

[a] Costs are in levelized 1980 $.
[b] Peak year is 2010.
[c] Peak year is 2020.

cumulative emission reduction over the 50-yr time horizon (together with cumulative cost). Cumulative reduction could be an important measure of effectiveness of a strategy if, for example, scientific research determined that ecological damage could be minimized by reducing cumulative deposition over long time periods (if the damage mechanism was chronic or cumulative in nature). Table 7.4 shows several success measures for the four strategies discussed above, including cost-effectiveness. Careful examination of each attribute across scenarios should be a prerequisite for determining a preference.

The actual response of utility operators to requirements for reduced emissions is hard to predict. In some instances, inefficient, heavily emitting units may retire early and be replaced with either new coal capacity, nuclear capacity, or increased utilization of existing units. Acid rain legislation would certainly force utilities to reevaluate their mix of plants for system optimization, among new coal plants, nuclear or other nonpolluting plants, refurbished plants, old coal plants with and without FGD, etc. Their choices are hard to predict, and, therefore, future levels of emissions are also hard to predict.

REGULATORY EFFICIENCY

The existing regulatory program for controlling the emissions of acid-precursor pollutants from major sources derives its authority from the Clean Air Act as amended in 1977. The provisions of this act were generally designed to protect local air quality values--that is, to avoid deterioration of air quality caused by identifiable local sources rather than to protect regional air quality values, such as acid deposition or regional haze, where culpable sources cannot be readily identified. To accomplish this, emission limits for major existing sources were established to bring polluted areas into compliance with National Ambient Air Quality Standards (NAAQS) and to maintain compliance. Stringent emission limits, equivalent to the performance of best demonstrated technology, were placed on all new sources in the form of New Source Performance Standards (NSPS). The existing scheme of emission limitation seems adequate to assure acceptable air quality at the local level; and, to the extent that compliance has been achieved, the scheme has proved successful.

However, as local air quality gradually improves, but regional air quality remains unchecked, this philosophy may need rethinking. There is no *a priori* reason why the regulatory mechanism found suitable for local air pollution problems should be suitable for regional air pollution problems. In fact, there is good reason to believe that the status quo is a very inefficient approach to the regional air pollution problem. Provided no violations of NAAQS or other local air quality values are caused, methods of regional air quality protection that are more cost-effective should be considered, even if they require revisions to the Clean Air Act. The bubble policy may offer improved efficiency on both economic and environmental grounds, particularly if new and existing sources are permitted to bubble together.

The bubble policy is an environmental emissions management strategy that regulates the emissions from a group of point sources as if those emissions came from a single source. A conceptual "bubble" is placed over all the sources, and only the total emissions leaving the bubble are regulated. Individual firms can then choose the most cost-effective mix of control technologies at each plant or emitting source, under a single constraint on total emissions. Regulatory activity to date has centered on air pollution bubbles for industrial manufacturing plants and electric utility plants in proximate locations; however, regulatory bubbles have also been suggested for water pollutants (Star, 1982; Muffat, 1983; "EPA to Revise", 1983).

Bubbles that have met with U.S. Environmental Protection Agency (EPA) approval generally have regulated only existing sources; new sources within bubbles are still required to meet NSPS. These stricter controls on new sources are meant to ensure a natural evolution to a cleaner environment, even with economic growth, as new sources come on line and old sources retire. There is an increasing realization, however, that this evolution is slow and may actually be impeded by stringent NSPS requirements. Facility owners prefer to extend the lifetime of existing units rather than face the high capital costs of control technology for new units. Delayed environmental cleanup may be a perverse result of this strong enviornmental policy.

An additional reason for precluding bubbling among new and existing sources has been the congressional concern about regional competition for new sources. One intention of the NSPS was to forestall any attempts by regions or states to attract new industry by establishing less-stringent (and thus less-expensive) requirements for air pollution control. This concern may be valid for industrial manufacturing sectors, but it does not apply to electric utilities, which are largely confined to geographical regions by their nature and their charters, especially in the East.

FIGURE 7.10
The effects of a regional bubble policy on power plant emissions

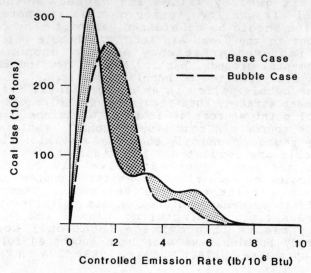

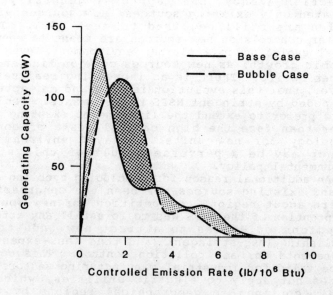

The EPA recognizes the potential cost savings that would result if new sources were allowed to bubble with existing sources. To avoid relaxing controls on new sources and forgoing benefits in the long term, EPA is considering the "declining bubble" concept, whereby allowable emissions would decline over time, as old sources retire, along the same emissions track as would occur under current regulations ("NSPS Bubble Issues", 1982). Other work has shown that a "regional declining bubble" that tracks four million tons below emissions under current regulations out to the year 2030 costs about the same as current regulations (Garvey, et al., 1984).

The use of bubble strategies has also been approved by EPA at the facility level, with the overall emissions constraint applying to all sources at a particular site or in a small locale. However, if certain air quality problems are regional in nature (e.g., long-range transport and regional haze) and the exact locations of the emissions sources are unimportant, then there may be advantages to extending the bubble beyond the boundaries of a single site to a much larger, regional scale.

The expansion of the bubble policy from the facility level to the regional level is not as great a step as it might seem; the projected cost savings would probably be nearly as great at the facility level. The reason for this similarity is that, if new sources are allowed to bubble with existing sources, they would in all likelihood colocate with those existing sources where the bubbling potential (and thus the potential for cost savings) is greatest. The benefit obtained would be suboptimal but probably not too dissimilar from the optimal benefit. The important assumption is that new and existing sources are allowed to bubble together.

Recent studies of regional, new-source/old-source bubble policies for coal-fired power plants in the East have shown that major benefits can be achieved (Streets, et al., October 1983; Streets, et al., 1984b). Benefits would arise through reducing emissions at old (SIP) plants while relaxing the stringent emission limits for new (NSPS) plants. Plants can find cost-effective solutions by emitting in the middle range of 2-3 lb $SO_2/10^6$ Btu, rather than at approximately 0.6 lb/10^6 Btu (under NSPS) or at rates as high as 6-7 lb/10^6 Btu (under SIPs). The changes in emission rates under a regional bubble policy are illustrated in Figure 7.10. The existing SIP-NSPS mechanism is not an efficient way to regulate total emissions from a group of sources.

Figure 7.11 shows various bubble options compared to the current regulatory scheme. The options are a

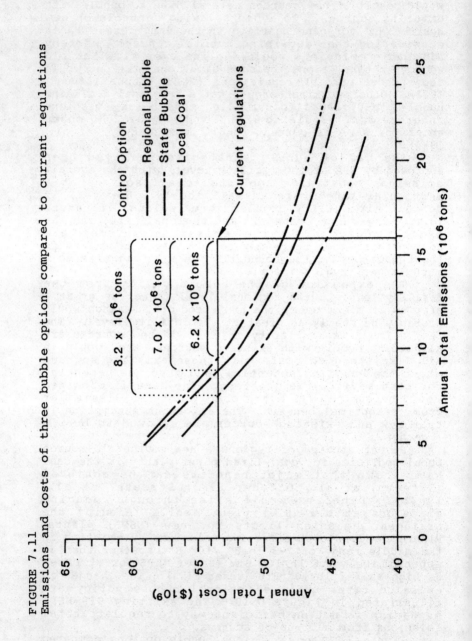

FIGURE 7.11
Emissions and costs of three bubble options compared to current regulations

regional bubble (in which all units in the eastern United States can participate), a statewide bubble (in which each state has its own bubble), and a statewide bubble in which the use of hitherto untapped reserves of medium-sulfur coal is hypothesized. In each case, all 1,333 existing and projected new coal-fired powerplant units (to the year 2000) are included in the analysis. Total costs presented include all the costs of pollution control, together with all fuel costs, such that the differences between scenario costs adequately reflect coal premiums due to fuel switching.

The current regulatory policy is represented by the single point corresponding to SO_2 emissions of 15.9 million tons/yr from all old plants under SIPs and new plants under NSPS, at a total cost of \$53.4 billion/yr. Each bubble option is expressed as a cost/emissions curve, for, clearly any desirable level of emissions can be specified for the group of plants under the bubble. Intersections of these curves with the vertical line corresponding to 15.9 million tons/yr represent the potential cost savings of the policies in achieving the same level of emissions as current regulations. Intersections of the curves with the horizontal line corresponding to \$53.4 billion/yr represent the potential emission reductions achievable at the same cost as the cost to comply with current regulations. Table 7.5 lists the potential cost and emission savings. Any points on the curves between these horizontal and vertical lines represent combined cost savings and emission reductions.

In the context of an acid deposition control program, these potential SO_2 emission reductions at no cost (6-8 million tons/yr) are substantial. they are close in magnitude to the reductions mandated by the bills (see Figure 7.4), which cost \$2-4 billion/yr by conventional control methods. Even a statewide bubble for the 1,172 existing sources could potentially achieve a 3.5 million tons/yr reduction in SO_2 emissions at no cost (Table 7.5). the conclusion that great inefficiencies exist in the current SIP-NSPS regulatory scheme to limit regional emissions is inescapable.

Implementation of a bubble policy on a regional scale presents difficulties in determining which sources should be controlled to achieve a desired level of emission. Ideally, sources would determine this among themselves to achieve greater efficiency. Naturally, the advantages of the bubble policy as an emission reduction strategy are subject to the same uncertainties and caveats as are the cost savings. However, these uncertainties take on greater significance when the bubble policy is viewed as an emission reduction strategy. If the bubble policy is

TABLE 7.5
Summary of potential benefits of alternative bubble options

Bubble Option	Number of Bubbled Units	Capacity of Bubbled Units (GW)	Potential Cost Savings[a] ($10^9/yr)	Potential SO_2 Emission Reductions[b] (10^6 tons/yr)
Regional, new and existing sources	1333	316	5.9	7.0
Statewide, new and existing sources	1333	316	4.8	6.2
Statewide, existing sources only	1172	228	1.8	3.6
Statewide, new and existing sources local coals utilized	1333	316	7.7	8.2

[a] To achieve the same level of SO_2 emissions as would occur under current regulations.
[b] At the same cost to sources as the costs to comply with current regulations.

instituted primarily to save money for utility companies, it will become a voluntary, self-governing policy. That is, utilities will use the policy if it is truly advantageous and ignore it if it is not, assuming that each utility wishes to minimize costs. If, however, the bubble policy is instituted to reduce emissions (of the magnitude indicated in Table 7.5), then compliance would have to be mandatory because utilities would have no financial incentive to change their existing operations. In this context, the possible uncertainties and implementation problems associated with the bubble become more critical.

REAL-WORLD EFFICIENCY

The preceding examples are all ways in which optimal configurations of emission reductions can be chosen to yield more-efficient approaches to solving the acid deposition problem than are contained in congressional bills. All of these solutions would be implemented through a mandatory emission reduction program run by state or federal government. The details of how emission rates would be determined for each plant under each of these programs remain to be specified. The flexibility given to individual or groups of sources could ultimately determine how well a hypothetically efficient control strategy translates into real-world efficiency. Economists would stress that real-world efficiency is more likely to be achieved through programs that offer economic incentives for or impose economic penalties on individual sources. If the sources make rational economic choices then the emission reduction goal should be achieved efficiently (Pechman, 1984; Proceedings, 1981).

Economic incentives or penalties can take various forms. Emission taxes can be set equal to the marginal cost of cleanup to a desired level. A system of marketable emission rights could be established. Each of these options would limit the role of government to setting up a system that offers a range of economic choices for source operators. There has been much talk of such strategies for controlling acidic deposition, but only one concrete proposal: a minor bill, H.R. 4483, introduced by Rep. Les Aspin on November 18, 1983, to amend both the Clean Air Act and the Internal Revenue Code of 1954. This interesting and unique proposal deserves close scrutiny.

House Bill H.R. 4483 was designed to establish a national program of tax incentives and financial assistance to reduce SO_2 emissions. It is unique in that it is the first acid rain bill to provide

FIGURE 7.12
Emissions and costs frontier for a typical utility plant in Illinois under the Aspin bill

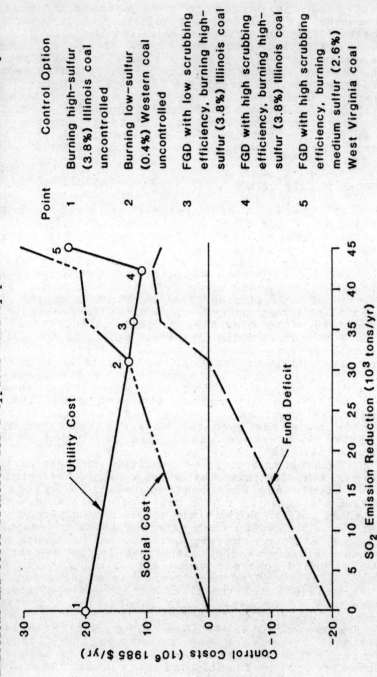

utilities with a program of economic incentives without mandated emission reductions. The Aspin bill would impose a tax of 50 cents for every pound of sulfur per ton of coal, to the extent that the coal exceeds 0.5% sulfur by weight (10 lb of sulfur per ton of coal). The utility would receive a rebate of 50 cents for every pound of sulfur removed from the coal by pre- or post-combustion techniques. To minimize disruptions in the coal mining industry and to provide additional economic incentives to control SO_2 emissions, H.R. 4483 would subsidize 90% of the capital expenditures for pre- and post-combustion sulfur removal systems for which construction commenced after enactment.

If enacted, the Aspin bill would significantly alter the economic choices facing the electric utility industry. The after-tax cost of some high-sulfur coals would more than double, while utility costs for flue-gas desulfurization systems (FGD) would be drastically reduced or would actually result in a cost savings. Argonne National Laboratory has analyzed the implications of this bill, assuming that utilities act in an economically efficient manner, each utility selecting the coal and SO_2 control technology combination that minimizes its total costs.

The Aspin bill provides utilities with economic stimuli for reducing SO_2 emissions by fuel switching or controlling SO_2 emissions. A utility may switch to a lower-sulfur coal and reduce sulfur tax expenditures or scrub a high-sulfur coal and receive a 90% subsidy for FGD capital costs and 50 cents for every pound of sulfur removed. In some cases a utility can minimize its costs by both fuel switching and scrubbing. The option chosen by a utility depends on factors such as the unit's age, capacity, geographical location, alternative fuel options, and utilization rate.

Figure 7.12 illustrates the effectiveness of the Aspin bill in establishing economic incentives for reducing SO_2 emissions. The figure shows SO_2 removal costs and emission reductions for various levels of SO_2 control for a typical large coal-fired unit in Illinois. The utility curve represents the cost frontier for the utility to achieve various levels of SO_2 emission reduction. Utility costs are expenditures above the status quo and include taxes paid to the fund, fuel premiums and incremental particulate control costs associated with fuel switching, 10% of FGD capital costs, FGD operation and maintenance costs, energy costs to power FGD systems, and a 50 cents credit for every pound of sulfur removed by FGDs. The social costs are the actual costs to society to implement the control option selected by the utility. Social costs include all expenditures associated with fuel switching and FGD systems.

Figure 7.12 shows that if the utility continues to burn high-sulfur Illinois coal without scrubbing (Point 1) it will pay $20 million/yr into the sulfur tax fund. The utility could lower its operating costs by switching to a lower-sulfur western coal (Point 2). The fuel premium paid for low-sulfur coal is less than the sulfur tax. Although the utility lowers its costs by fuel switching, the cost to society for the low-sulfur coal premium is $15 million/yr. The utility could decrease its operating costs further by scrubbing high-sulfur Illinois coal at a 75% removal efficiency (Point 3). At this point the utility receives more money from the fund for the 90% capital subsidy and 50 cents/lb sulfur rebate than it pays into the fund. The utility minimizes its costs by controlling SO_2 emissions to the point where the marginal cost of increasing the scrubber removal efficiency is equal to the incremental revenue received from the fund (Point 4). Utility operating costs rapidly increase when SO_2 emissions are controlled further (Point 5). This example illustrates the method by which the Aspin bill establishes cost incentives for utilities to reduce their emissions.

Algorithms were developed to calculate utility costs, total social costs, and fund deficit. These were applied to each power-plant unit in the nation, assuming that utilities would apply the level of control that minimizes costs to themselves. On this basis it was determined that 6.5 million tons of SO_2 would be reduced at a total social cost of $2.6 billion/yr. Figure 7.4 shows how this compares with other mandatory emission reduction bills in terms of economic efficiency relative to a least-cost strategy. Much of this reduction would be achieved by installing 55 GW of scrubbing capacity. Approximately 80% of the SO_2 reduction would occur in eight states: Georgia, Illinois, Indiana, Missouri, Ohio, Pennsylvania, West Virginia, and Tennessee. The Aspin bill would have little effect on western states where utilities primarily burn low-sulfur coal.

The sulfur tax paid by the utilities is expected to generate enough funds to pay for almost all of the total cost of implementing H.R. 4483 in the early years of the program. Although there is a small deficit of $14 million in 1985, the deficit is expected to increase in the future as older units retire. Much of the fund's revenue comes from old units that cannot economically install FGD systems or switch to a low-sulfur coal. As old units retire the revenue generated from the sulfur tax will diminish. Newer units generally install FGD systems and receive more money from the fund than they pay in taxes. Section 183(c) authorizes the fund administrator to make expenditures in accordance with a priority list if funds are inadequate. The priority list would be based on the dates on which applications for subsidy were approved. This provision could act as an incentive to plants to participate early.

The Aspin bill is unique in its approach to emission reductions for acid rain control. It is unlikely to be adopted because of problems associated with the range of sources that would be included; plants required to install control systems under NSPS would appear to be eligible for subsidies. However, one might hope that it would act as a springboard for more refined proposals based on the same principle. Control strategies that place options squarely in the hands of source owners and operators may truly be the ones that maximize efficiency.

ACKNOWLEDGMENTS

Argonne National Laboratory (ANL) and the U.S. Department of Energy (DOE) have been involved in an interactive program of environmental analysis for almost ten years. Strategies to reduce emissions of air pollutants in environmentally and economically sound ways have been a major focus of attention for much of this time. It gives me pleasure to acknowledge the analytical assistance and ideas provided by my colleagues: Doug Carter, Ed Trexler, Jean Vernet, and Ted Williams (all DOE); and Les Conley, Doris Garvey, Paul Grogan, Don Hanson, Duane Knudson, Marylynn Placet, John Roberts, and Tom Veselka (all ANL). Without their good sense and enthusiasm, the work described in this chapter could not have been performed. Funding was provided by DOE's Office of Policy, Safety, and Environment (formerly the Office of Environmental Protection, Safety, and Emergency Preparedness), and DOE's Office of Fossil Energy. The opinions expressed herein should not be construed as supportive of any policy of Argonne National Laboratory or the U.S. Department of Energy.

REFERENCES

Bullard, C.W., and H. Hottmann. "Strategies for Reducing Acid Emissions From Illinois Electric Utilities: A Preliminary Assessment." A paper prepared for the State of Illinois: Energy Resources Commission and Environmental Protection Agency, September 1984.

Damon, J.E., P.A. Ireland, D.M. Shattuck, E.W. Stenby,

and D.G. Neumeyer. "An Approach for Selecting Cost Effective Acid Rain Control Technologies." Presented at the 1984 American Power Conference, Chicago, Ill., April 1984.

E.H. Pechan & Associates, Inc. "AIRCOST Model: Technical Documentation." Springfield, VA.: E.H. Pechan & Associates, April 1983.

Energy Information Administration. "Impacts of the Proposed Clean Air Act Amendments of 1982 on the Coal and Electric Utility Industries." [DOE/EIA-0407] Washington, D.C.: U.S. Department of Energy, June 1983.

Energy Security Act of 1980, Title VII (P.L. 96-294).

"EPA to Revise Steel Water Bubble to Require Net Reductions in Pollution." **Inside EPA Weekly Report.** March 1983.

Fay, J.A., D. Golomb, and J. Gruhl. "Controlling Acid Rain." [Report No. MIT-EL 83-004]. Cambridge, Mass.: M.I.T. Energy Laboratory, April 1983.

Ferguson, H.L. and L. Machta. Final Report of Work Group 2 submitted to the Coordinating Committee created under the August 5, 1980, Memorandum of Intent on Transboundary Air Pollution between the United States and Canada, November 1982.

Garvey, D.B., D.G. Streets, and T.D. Veselka. Unpublished information. Argonne National Laboratory, May 1984.

ICF Inc. "Analysis of a Senate Emission Reduction Bill (S-3041)." Prepared for U.S. EPA. Washington, D.C.: EPA, February 1983.

ICF Inc. "Analysis of Cost-Effective, Phased-In Reductions of Sulfur Dioxide Emissions." Prepared for the Alliance for Clean Energy. Washington, D.C.: Alliance for Clean Energy, February 1984.

Keelin, T.W., and E.N. Oatman. **Public Utilities Fortnightly.** December 1982.

Muffat, J.C. "Start the Bubble Machine." **Environmental Progress.** 2, (1983).

National Research Council. **Atmosphere-Biosphere Interactions: Toward a Better Understanding of the Ecological Consequences of Fossil Fuel Combustion.** Washington, D.C.: National Academy Press, 1981.

"NSPS Bubble Issues." Prepared by EPA for discussion at the National Air Pollution Control Techniques Advisory Committee Meeting, June 1982.

Office of Technology Assessment. "An Analysis of the Sikorski/Waxman Acid Rain Control Proposal: H.R. 3400, 'The National Acid Deposition Control Act of 1983.'" Washington, D.C.: U.S. Congress, Office of Technology Assessment, July 1983.

Parker, L.B., and L. Kumins. "Sharing the Cost of Acid

Rain Control: An Analysis of Federal Financing Under H.R. 3400." Congressional Research Service, Washington, D.E.: C.R.S., September 1983.

Pechman, C. "Equity, Efficiency and Sulfur Emission Reductions." A paper (84-27.2) presented at the 77th Annual Meeting of the Air Pollution Control Association, San Francisco, Calif., June 1984.

Proceedings of the Special Conference on Economic Incentives for Clean Air, Air Pollution Control Association, San Francisco, Calif., January 1981.

Schwengels, P., E. Pechan, K. Graves, and D. Streets. "An Integrated Emissions MOdel System for Analysis of Acid Deposition Control Strategies." A paper (84-27.1) presented at the 77th Annual Meeting of the Air Pollution Control Association, San Francisco, Calif., June 1984.

Shannon, J.D. **Atmos. Environ.**, 1 (1981).

Star, A.M. "Emissions Trading Offers New Choices for Pollution Control." **Power Magazine.** December 1982.

Streets, D.G., L.A. Conley, L.D. Carter, and J.E. Vernet. "An Analysis of Proposed Legislation to Control Acid Rain." [ANL/EES-TM-209]. Argonne, Ill.: Argonne National Laboratory, January 1983.

Streets, D.G., D.B. Garvey, P.J. Grogan, D.A. Hanson, and L.D. Carter. **J. Air Pollut. Control Assoc**, 34 (1984b).

Streets, D.G., P.J. Grogan, D.A. Hanson, and T.D. Veselka. "A Regional New-Source Bubble Policy for Sulfur Dioxide Emissions in the Eastern United States." [Report ANL/EES-TM-239]. Argone, Ill.: Argonne National Laboratory, October 1983.

Streets, D.G., D.A. Hanson, and L.D. Carter. **J. Air Pollut. Control Assoc.** 34, 1984a.

Stukel, J.J., and C.W. Bullard. "Strategies for Reducing Sulfur Dioxide Emissions From Midwest Electric Utilities." A paper 84-27.4 presented at the 77th Annual Meeting of the Air Pollution Control Association, San Francisco, Calif., June 1984.

Systems Applications, Inc., and ICF Inc. "A Study of Acid Deposition Reduction Costs." Prepared for Public Service Indiana, August 1983.

Teknekron Research Inc. "Electric Utility Emissions: Control Strategies and Costs." [Report No. RK-80-2057]. Berkeley, Calif.: Teknekron Research, Inc., April 1981.

Temple, Barker & Sloane Inc. "Evaluation of H.R. 3400, the 'Sikorski/Waxman' Bill for Acid Rain Abatement." Washington, D.C.: Edison Electric Institute, January 1984.

Trisko, E.M. "Acid Precipitation: Causes, Consequences, Controls." **Public Utilities Fortnightly.** February

1983.
Universities Research Group on Energy. "The State-Level Advanced Utility Simulation Model: Analytical Documentation." Urbana, Ill.: University of Illinois at Urbana-Champaign, September 1983.

8
Equitably Reducing Transboundary Causes of Acid Rain: An Economic Incentive Regulatory Approach

David J. Webber

INTRODUCTION

One dimension of the current public discussion of acid rain is that of the equity of one state, region, or nation imposing environmental damage on another through the emission of airborne pollutants. Both principles of international law, as set forth in the 1972 Declaration of the United Nations Conference on the Human Environment, and widely-held notions of equity and fairness, dictate that transboundary pollution be corrected in as equitable and speedy a fashion as possible. Remedial policy action, on the other hand, has not been forthcoming due to the political infeasibility of requiring the pollution producing jurisdiction to bear the economic burden of reducing its industrial and vehicular emissions.
Generally, the consensus among environmental advocates suggests that environmental quality has improved over the past decade. For example, The Conservation Foundation's assessment of the condition of the environment at mid-decade states:

> The United States has made significant progress in many, if not all, environmental areas where laws and institutions have been explicitly devised to address specific problems. Most of the conventional air pollutions no longer pose a health threat to nearby communities; the majority of the nation's rivers are suitable for fishing and swimming; exposure to some specific toxic substances--lead, PCBs, chlorinated hydrocarbon pesticides such as DDT--had declined; the populations of many wildlife species are increasing (The Conservation Foundation, 1984: 1).

Likewise, the National Wildlife Federation, which has

calculated an Environmental Quality Index for over a decade, reports that, based on a subjective evaluation of seven environmental resources, three (water, living space, and soil) are declining in quality and four (wildlife, air, minerals, and forests) have stabilized (National Wildlife, 1983).

On the other hand, environmental policies have been criticized for being ineffective, for contributing to inflation and poor industrial productivity, and for being an unjustified intrusion into private concerns and local or state governmental activity. Environmental policy is currently in a period of reconsideration. The basis for this reconsideration is aptly described by Ingram and Mann:

> Environmental policy has moved from an era of commitment to environmental quality goals to a period of searching for efficient, economical, and politically feasible techniques and mechanisms for protecting the environment. In the early 1970s, far-reaching pollution control legislation swept through Congress on the wave of public recognition of environmental degradation and public belief that the government had done not nearly enough to halt pollution. As the decade progressed, and the immense governmental task of administering ambitious legislation have become increasingly salient, shortages, economic dislocation, and administrative conflicts have led to questions about whether too much is being attempted too fast and in ineffective ways (Ingram and Mann, 1978: 131).

Acid rain reduction proposals must be developed and considered in light of this cautious, economically-minded political environment.

One way that progress can be made in reducing acid rain is by the adoption of an economic-based incentive regulatory scheme that assigns responsibility for controlling sulfur and nitrogen emissions causing acid rain to the jurisdiction of origin. In the context of the national acid rain controversy, the responsible jurisdiction would be states or air quality regions; for the United States-Canadian transboundary problem, responsibility would belong to the appropriate national government.

In addition to the widely-recognized economic advantages of an economic incentive regulatory approach (i.e., economic efficiency, production flexibility, and reduced administrative costs), economic incentives designed to reduce acid rain are attractive for several other reasons. Among these are: equity, administrative flexibility, and ease of implementation. The latter includes the possibility that economic-incentive regulation can be phased in over a period of time, thereby allowing for alternative corrective action to be undertaken. Despite these attractive features economic incentives are not a miracle cure for acid rain (or any other policy problem). A properly designed incentive scheme requires volumes of information about sources and consequences of various types of emissions, correctly assigned incentive fees or subsidies, and careful monitoring of emissions.

This chapter proposes an economic-based incentive regulatory scheme intended to reduce sulfur oxide emissions. This analysis is presented in four sections. First, acid rain and the regional distribution of sulfur oxide producers are briefly reviewed. Included in this section is a review of the state of atmospheric modeling in identifying and tracing sulfur oxide emissions. The next section discusses the current regulatory procedures for promoting clean air by the control of specific air emissions. The advantages of and requirements for an economic incentive approach to designing environmental regulations are then discussed. Finally, the last section outlines an economic incentive strategy for controlling sulfur oxide causes of acid rain and examines how this proposal can be incorporated into the current structure of the air pollution control program outlined in the state implementation provision of the Clean Air Act.

THE REGIONAL DISTRIBUTION OF CAUSES AND EFFECTS OF ACID RAIN

Acid rain, the commonly used term for what is more properly called "acid deposition," is the result of sulfur and nitrogen oxides reacting with water vapor in the atmosphere and is usually desposited to the earth in rain or snow. Since pure rain has a natural pH of 5.6, precipitation with a pH of less than 5.6 indicates the presence of some acidic substance. Precipitation in the eastern United States frequently has a pH in the 4.0 to 5.0 range with a low of 1.4 once reported for Wheeling, West Virginia.

Acid rain has become an issue of public concern over the past five years or so because of its potential for damage to forests, lakes and streams, crops, soil fertility, fish, and manmade materials. Lake acidification in the northeastern United States is perhaps the single best manifestation of the acid rain problem. As acidic rain falls on lakes and streams and

TABLE 8.1
1980 state suflur dioxide and
nitrogen oxide emissions by
state (1,000 tons/year)

State	SO_2	NO_x
Alabama	760	450
Alaska	20	55
Arizona	900	260
Arkansas	100	220
California	445	1,220
Colorado	130	275
Connecticut	70	135
Delaware	110	50
District of Columbia	15	20
Florida	1,100	650
Georgia	840	490
Hawaii	60	45
Idaho	50	80
Illinois	1,470	1,000
Indiana	2,000	770
Iowa	330	320
Kansas	220	440
Kentucky	1,120	530
Louisiana	300	930
Maine	100	60
Maryland	340	250
Massachusetts	340	250
Michigan	900	690
Minnesota	260	370
Mississippi	280	280
Missouri	1,300	570
Montana	160	120
Nebraska	75	190
Nevada	240	80
New Hampshire	90	60
New Jersey	280	400
New Mexico	270	290
New York	950	680
North Carolina	600	540
North Dakota	100	120
Ohio	2,650	1,140
Oklahoma	120	520
Oregon	60	200
Pennsylvania	2,020	1,040
Rhode Island	15	50
South Carolina	330	200
South Dakota	40	90
Tennessee	1,100	520
Texas	1,270	2,540
Utah	70	190
Vermont	7	25
Virginia	360	400
Washington	270	290
West Virginia	1,100	450
Wisconsin	640	420
Wyoming	180	260
U.S. TOTAL	26,500	21,220

Source: Office of Technology Assessment, Acid Rain and Transported Air Pollutants, OTA-)-204, June 1984, p. 58

their surrounding environs, a complex reaction is set off that can ultimately result in the reduction of plant and fish life that can mean both the destruction of a natural ecology as well as the loss of fishing, farming, forestry, and tourist industries.

The major substances contributing to acid rain are sulfuric and nitric acid, with the former accounting for about 70 percent of acid rain and the latter accounting for the remaining 30 percent. While there is certainly scientific disagreement, the principal precursor pollutants resulting in acid rain are produced chiefly by the combustion of fossil fuels in powerplants, industrial boilers, and automobiles. The single greatest source of acid rain is believed to be the sulfuric acid that is the product of chemical reactions of the sulfur dioxide resulting from the combustion of sulfur in the coal burned by electric utilities.

Table 8.1 reports the 1980 estimated sulfur dioxide and nitrogen oxide emissions by state. Of the 26,500 thousand tons of sulfur dioxides emitted in 1980, approximately 21 thousand tons, or 80 percent, were emitted by the 31 eastern states. Ohio, with 2,650 thousand tons, has the highest emissions followed by Pennsylvania (2,020), Indiana (2,000), Illinois (1,470), Missouri (1,300), Texas (1,270), Kentucky (1,120), and Tennessee and West Virginia with 1,100 thousand tons each (OTA, 1984: 57-58). Electric utilities are responsible for an estimated 17 million tons of sulfur dioxide emissions, with industrial boilers and smelters accounting for about 7 milion (OTA, 1984: 60).

Nitrogen oxides emissions, estimated to be 21,200 tons in 1980, are more evenly distributed across the states as shown in Table 8.1. The eastern 31 states are responsible for about 14,000 tons, or two-thirds, of the national nitrogen oxide emissions. Mobile sources are responsible for 9,400 tons and utilities are the source of approximately 6,400 thousand tons. The rapid growth in utility nitrogen oxide emissions is expected to make them the largest single source of nitrogen oxides by the year 2010 (OTA, 1984: 51).

While projecting the magnitude of future emissions is a difficult task requiring assumptions about future demand for electricity, the retirement rate of existing emission sources, and the effectiveness of environmental regulations, the Office of Technology Assessment (OTA) of the United States Congress estimates that sulfur dioxide emissions will increase by 10 to 25 percent by the year 2010 with the most rapid increase coming in the industrial sector. Compared with the breakdown of 1980 sulfur dioxide emissions, which were divided 65 percent from utilities and 28 percent from industry, the 29.3 thousand tons estimated to be emitted in the year 2000 are expected to be divided 61 percent from utilities and 32 percent from industrial sources (OTA, 1984: 59).

Nitrogen oxide emissions are expected to increase to 26.6 million tons with the largest increase coming from utilities. Compared to the 1980 breakdown of 44 percent of emissions coming from mobile sources and 30 percent from utilities, 40 percent of nitrogen oxide emissions in 2000 will result from mobile sources and 36 percent from utilities (OTA, 1984: 59-60).

While environmental damage is widely thought of as being isolated to the northeastern United States, several areas of the eastern and midwest regions of the country are susceptible to the risks of acid precipitation. The largest number of lakes and streams that are vulnerable to acidification are found in Minnesota and Wisconsin, followed by the New England states, and the mountain states of West Virginia, Kentucky, North Carolina, and parts of Virginia and Tennessee (OTA, 1984: 80-81).

Although the Adirondack Mountains in New York are the only site in the United States where declines in fish population have been documented, considerable economic risk is faced by several regions of the country due to their exposure to acid rain and to the sensitivity of their aquatic systems. These regions include the northern New England States (Maine, Vermont, and New Hampshire), the Appalachian region (West Virginia and eastern Kentucky), and the midwest (parts of Wisconsin and Minnesota (OTA, 1984: 82).

Agricultural and forest declines due to acid rain are also distributed outside the New England region. Several studies show reductions in soybeans and corn production in Illinois, Indiana, and Ohio and growth declines in several tree species have been noted as far south as Georgia (OTA, 1984: 82-85). In fact, many believe that it is the potential damage to forests that is the greatest harm from acid rain.

Sulfur dioxide emissions also create a health risk which is widely distributed across the states. The high population density states of Ohio, Illinois, Indiana, and Pennsylvania are thought to be at greatest risk but Michigan, Texas, North Carolina, and Virginia as well as the New England states also face high health risks associated with sulfate exposure (OTA, 1984: 91-92).

THE CURRENT AIR QUALITY CONTROL REGULATORY SCHEME AND ACID RAIN PROPOSAL

The Clean Air Act of 1970, as amended several

times since, is the nation's major statute outlining an air quality control program. The act is generally seen as incapable of correcting the current acid rain problem (see Gallogly, 1981) and is often criticized for its administrative complexity. For example, Crandall describes the Act as follows:

> The Clean Air Act Amendments of 1970 created a curious variety of federalism. State governments set individual standards for existing polluters, but the federal government dictates the goals. New-source standards are issued by the EPA, but offsets of emissions from older sources against emissions from new sources are administered by the states. Most enforcement responsibilities reside with the states, but the federal government retains civil penalty powers for punishing violators. The EPA can declare an air quality control region in nonattainment for one or more pollutants, but it can force compliance only through indirect incentives such as withholding of federal grant monies (Crandall, 1983: 8).

The principal framework of the Clean Air Act is the establishment of "national ambient air quality standards" (NAAQS) in order "to protect the public health and welfare." These standards are set by the EPA and each state is required to submit a State Implementation Plan (SIP) showing how the standards will be implemented and enforced in that state. Therefore, a uniform national air quality standard is adopted but individual states have discretion in regulating their own emissions in order to achieve that standard.

This regulatory scheme has not been effective in dealing with the sulfur dioxide causes of acid rain because SIPs are focused on the emissions from a state's emission sources as measured at a monitoring station located within that state. Sulfur dioxide emitted high in the atmosphere will most likely satisfy the provisions of the state's implementation plan but will be deposited in another state up to several hundred miles away. The construction of "tall stacks" has been encouraged by this "monitoring within the state" view of the NAAQS as carried out by the SIP's. Because the Clean Air Act gives each state the primary responsibility for the control of emissions (from old sources) to the state within which the source is located, interstate abatement is not likely to occur. And, while there are several interstate provisions of the Clean Air Act, most notably section 115, the Act

TABLE 8.2
Top 50 plants by state

State	Number of Plants	Number of Units	1980 Annual SO_2 Emissions (Thousands of Tons)
Alabama	1	5	122
Florida	3	12	280
Georgia	2	9	307
Illinois	4	13	722
Indiana	7	26	1,025
Kentucky	4	11	576
Mississippi	1	5	62
Missouri	5	13	889
Ohio	9	49	1,594
Pennsylvania	4	11	675
Tennessee	5	23	779
West Virginia	4	10	664
Wisconsin	1	8	120
TOTAL	50	195	7,815

Source: Edison Electric Institute, "An Evaluation of H.R. 3400: Sikorski/Waxman Bill for Acid Rain Abatement," p. 11.

was not designed with complex problems like the transboundary transfer of acid rain precursors in mind.
Because of the inability of the current Clean Air Act to correct the acid rain problem, several proposals are currently being considered in Congress as it prepares for the extension of the Act. The major acid rain conrol bills all focus on a sulfur dioxide reduction. The bills have a wide variety of approaches to achieve the intended reduction and differ in the size of the sulfur dioxide emissions prescribed, the way such rollbacks should be achieved, the way the cost burden will be distributed, and the geographic area that will be included in such sulfur reduction efforts.
The legislative proposal receiving the most public attention is H.R. 3400 introduced by Representatives Waxman and Sikorski and cosponsored by over 80 other House members. While this bill was defeated in the House subcommittee in May 1984, it served as the major vehicle for legislative efforts to reduce acid rain and contains the elements most likely to be included in future bills. The Waxman-Sikorski bill required a 10 million ton reduction (about 38 percent of 1980 emissions) of sulfur dioxide emissions. The 10 million ton reduction is to be achieved by installing scrubbing technology on the 50 largest utility emitters so that 90 percent of sulfur dioxide emissions are removed and by requiring the states (via their State Implementation Plans) to reduce emissions by about 3.5 million tons. H.R. 3400 provides for a 1 mill per kwh tax on all electricity generated in the United States or imported into the United States. The funds collected by this fee would be used to fund 90 percent of the mandated pollution control equipment on the 50 largest emitters (Congressional Research Service, 1984: 11).
As indicated in Table 8.2, Edison Electric Institute's list of the top 50 sulfur dioxide emitters includes plants that are located in 13 states with a combined 1980 emission level of 7,815 thousand tons. Of these 13 states, Ohio has the most targeted plants (9) followed by Indiana (7), Missouri (5), Tennessee (5), and Illinois, Kentucky, Pennsylvania, and West Virginia with 4 plants each (Edison Electric Institute, 1983).
The second specific provision of H.R. 3400 resulting in emission rollbacks is the reduction of additional emissions by an amount necessary to meet the 10 million ton reduction after the top 50 emitters have reduced. Table 8.3 reports how the additional 3.9 million ton reduction will be achieved by requiring reductions at the state level. As a percentage of the total 10 million ton reduction, Ohio will bear the largest responsibility with its rollback being 13.74 percent of the national total followed by Indiana

TABLE 8.3
H.R. 3400 required sulfur dioxide
emissions by state

	Total Reduction (Tons)	Top 50 Emitters Reduction (Tons)	Remaining Reduction	Percent of National Total
Alabama	283.64	74.90	208.74	5.49
Arizona	0.00	0.00	0.00	0.00
Arkansas	3.49	0.00	3.49	0.09
California	0.00	0.00	0.00	0.00
Colorado	0.30	0.00	0.30	0.01
Connecticut	0.00	0.00	0.00	0.00
Delaware	15.18	0.00	15.18	0.40
District of Columbia	0.00	0.00	0.00	0.00
Florida	322.92	248.90	74.02	1.95
Georgia	407.30	257.60	149.70	3.93
Idaho	0.00	0.00	0.00	0.00
Illinois	775.14	610.70	164.44	4.32
Indiana	1119.20	784.40	334.80	8.80
Iowa	120.60	0.00	120.60	3.17
Kansas	75.00	0.00	75.00	1.97
Kentucky	717.06	472.8-	244.26	6.42
Louisiana	4.90	0.00	4.90	0.13
Maine	3.07	0.00	3.07	0.08
Maryland	83.23	0.00	83.23	2.19
Massachusetts	61.25	0.00	61.25	1.61
Michigan	244.20	0.00	244.20	6.42
Minnesota	54.93	0.00	54.93	1.44
Mississippi	67.57	55.90	11.67	0.31
Missouri	950.78	735.60	215.18	5.66
Montana	1.50	0.00	1.50	0.04
Nebraska	7.50	0.00	7.50	0.20
Nevada	0.00	0.00	0.00	0.00
New Hampshire	46.14	0.00	46.14	1.21
New Jersey	45.52	0.00	45.52	1.20
New Mexico	0.00	0.00	0.00	0.00
New York	204.28	0.00	204.28	5.37
North Carolina	84.83	0.00	84.83	2.23
North Dakota	12.00	0.00	12.00	0.32
Ohio	1643.21	1120.30	522.91	13.74
Oklahoma	0.00	0.00	0.00	0.00
Oregon	0.00	0.00	0.00	0.00
Pennsylvania	827.98	607.90	220.08	5.78
Rhode Island	0.00	0.00	0.00	0.00
South Carolina	72.67	0.00	72.67	1.91
South Dakota	9.00	0.00	9.00	0.24
Tennessee	658.59	520.00	138.59	3.64
Texas	12.00	0.00	12.00	0.32
Utah	0.00	0.00	0.00	0.00
Vermont	0.00	0.00	0.00	0.00
Virginia	26.87	0.00	26.87	0.71
Washington	22.00	0.00	22.00	0.58
West Virginia	579.47	539.10	40.37	1.06
Wisconsin	322.38	60.10	262.28	6.89
Wyoming	7.30	0.00	7.30	0.19
TOTAL	9893.00	6088.20	3804.80	

Source: Edison Electric Institute, "Evaluation of H.R. 3400, p. 24.

(8.8%), Wisconsin (6.89%), Kentucky and Michigan (both with 6.42%) and Pennsylvania (5.78%).

Three variants of the Waxman-Sikorski bill are of special interest in this analysis because they relate directly to using economic incentives to reduce sulfur dioxide emissions. H.R. 4483, introduced by Congressman Aspin, establishes an "Acid Rain Deposition Control Fund" that would be collected by a 50-cent-per-pound tax on the sulfur content of coal exceeding 10 pounds of sulfur per ton. The tax would apply to both utility and industrial coal users and would be used to reimburse coal burning facilities 50 cents for each pound of sulfur removed from their emissions (Congressional Reference Service, 1984: 12).

Senate Bill 2001, introduced by Senator Durenberger, explicitly includes an emissions tax on both sulfur dioxide and nitrogen oxides. S. 2001 would raise $40 billion over a 10-year period to reimburse partially both capital and operating expenses for emission reductions at priority projects that would be designated by the Environmental Protection Agency. The Treasury Department is instructed in the bill to raise the $40 billion by placing an emissions tax sufficient to raise two-thirds of the fund from stationary sulfur dioxide sources, one-sixth from stationary nitrogen oxide sources, and one-sixth from mobile nitrogen oxide sources. The bill also requires a 10 million ton reduction in the 31 eastern states and caps sulfur dioxide emissions at 16.6 million tons by 1993 with offsets required for increases in emissions over the 1980 emission levels at individual stationary sources (Congressional Research Service, 1984: 12).

A third bill similar to the Waxman-Sikorski proposal is H.R. 4906 introduced by Representative Rinaldo. While requiring a 10 million ton reduction, this bill would allow states to use a capital expense fund to reimburse any technological equipment purchases in that state that reduce emissions. This fund is to come from a 1.5 mil tax on electricity generated from fossil fuel plants and will be in effect for only six years. The top 50 emitters are not singled out in this proposal and the individual states are allowed to determine where to make the necessary sulfur dioxide emission reductions (Congressional Reference Service, 1984: 12).

ADVANTAGES AND FEASIBILITY OF ECONOMIC-BASED ENVIRONMENTAL POLICY TOOLS

Over the past two decades a rather large literature has accumulated designing and proposing the use of economic incentives to encourage environmentally sound behavior (see Anderson et. al., 1977; Baumol and Oates, 1979; and Crandall, 1983). These market-based incentives are promoted as an alternative to direct governmental regulation and are designed to be more efficient and administratively easier to implement than the traditional promulgation and inspection type of regulation. Despite their theoretical elegance, the enthusiasm of economists and environmental policy analysts for economic-based policy tools has not been matched by that of policymakers in general. Recently, however, there appears to be a rejuvenation of interest in market-based incentives as indicated by the Environmental Protection Agency's initiation of emission offset programs and the consideration of the bubble concept for regulating air emissions in a geographic region (Blackman and Baumol, 1980; and Crandall, 1983).

The logic underlying economic-based policy tools is quite simple: since prices allocate scarce privately-owned resources, they can also be used to allocate scarce publicly-owned resources. After ownership rights have been established, producers of externalities should be charged the value of any third party effects generated. Theoretically, a firm generating environmental damage should be billed the total social costs of that damage such that the social resources affected are paid for by the user. The tax, fee, charge, or levy, therefore, forces the producers of externalities to consider social costs in making production decisions.

While the calculation of the value of social resources affected by a producer is very difficult in practice, the logic is clear: the price charged should cover the social cost of damage to the publicly-owned resources. In simple cases of air and water pollution it is relatively easy to estimate the appropriate price, but in most realistic environmental problem situations such calculations are very complex. For example, plant emissions into air and watersheds can be monitored and measured rather accurately and a per unit emission charge can be levied on the producer. The social costs of the discharge from the Reserve Mining Corporation in Minnesota into Lake Superior were estimated as stemming mainly from lake eutrophication, the increased need for water filtration, and the creation of a health hazard (Peterson, 1977). Such social cost estimates, which can serve as the basis for an effluent charge-based regulation scheme are relatively easy to calculate because they focus on damages to a semi-closed system--Lake Superior. On the other hand, the appropriate fee to be charged a producer of hazardous waste is difficult to conceptualize and even more difficult to calculate.

There are two difficulties incurred in approximating social valuations of environmental goods. First, public goods like the environment are so diffuse and amorphous that they are difficult to confine for description. Landscapes and scenic vistas, for example, are not easily described to citizens or policymakers asked to valuate them. Second, measuring collective preferences has proven quite challenging. While it is generally agreed by economists and other policy analysts that willingness to pay should be the central concept in calculating value, it is difficult to design effective estimation techniques. While the preferred method of cost estimation is a collective valuation of the foregone benefit as a result of the externalities produced, a simplified method focuses on the clean-up costs of the emissions and then allocates the total cost proportionately to each emitter. In this alternative method, willingness to pay and, therefore, social valuation of an environmental resource is implicitly included in a decision to expend scarce resources to clean up environmental damages, but there is no need to calculate the social value of the resource that is to be protected.

Internalizing externalities promotes the general social welfare in several ways. First, it promotes the efficient use of resources because it insures that all of the input prices are covered by the price of the finished product. Therefore, society can freely choose between the fully-priced product and a competing product on equal grounds. One good is not underpriced so that a disproportionate demand for that good might result in additional production of the socially wasteful good. Second, since all factors of production, including the external effects, are included, the efficient use of environmental resources is enhanced. Third, the social valuation of environmental goods can be incorporated into a producer's decision because the additional cost, the "pollution" tax or fee, that the producer must incorporate into future prices reflects the value of the disrupted environment. Fourth, while not required on efficiency grounds, the revenue generated by the pollution fee or tax is available for restoring environmental resources previously damaged.

Despite the advantages of economic efficiency offered by economic incentive-based policy tools, political support for such proposals has not been demonstrated. The incorporation of economic-based policy tools like the bubble concept and offsets into EPA's enforcement of the Clean Air Act has been a gradual process without widespread political support. Among the reasons offered for the lack of public support for such promising policy alternatives are the bureaucratic difficulties involved in monitoring emissions and setting an appropriate emission level and tax, ideological and ethical opposition to the de facto recognition of the inevitability of accepting a certain level of polluting emissions, and the inertia of the policy process to consider alternative regulatory schemes. Kelman (1983) questioned congressional staff members and lobbyists about their attitudes toward economic incentives and found that the most common reason for opposing a more widespread use of economic incentives in environmental regulation is ideological rather than disagreement with the effectiveness and efficiency of such regulatory schemes. Kelman finds that both supporters and opponents of market incentives are not well informed about the economic efficiency arguments surrounding economic incentives. Instead, their evaluations were likely to be consistent with their general ideological evaluation of the market system: Republican staffers tended to support proposals calling for the use of economic incentives while Democrats tended to oppose incorporating economic incentives into environmental regulations because of their distrust of the market system (Kelman, 1983: 292-293).

A STATE-BASED ECONOMIC INCENTIVE SO2 EMISSION CONTROL PROPOSAL

Because of the uneven distribution of causes, effects, and policy consequences of acid rain and acid rain abatement proposals, a traditional regulatory approach involving nationally uniform emission standards is not likely to be politically feasible or economically efficient. Simply put, the political feasibility of abatement proposals depends on designing a correction program that reconciles the economic interests of the major sulfur dioxide and coal producing states with the environmental interests of the states facing immediate and long term risks due to acid rain damage. The simple proposal of requiring sulfur dioxide emitters to "internalize externalities" by including social costs as a production cost covered by utility prices is unlikely to attract the necessary political support because of sunken costs, equity, and regional self-interest considerations.

Similarly, a stringent requirement of uniform emission reduction is not likely to be economically efficient in that there are a variety of ways that sulfur dioxide emitters might reduce emissions with lower economic and social disruptions. Uniform emission rollbacks and mandatory scrubbing regulations do not provide the managerial and administrative flexibility that might achieve a reduction in sulfur dioxide emissions at lower aggregate costs.

A state-based emissions reduction program relying heavily on a quasi-market regulatory scheme can reduce sulfur dioxide emissions in a way that recognizes regional diversity and interests. Such a state-based regulatory scheme requires the following:

1. A ceiling or maximum amount of sulfur emissions that each state can emit;

2. A maximum amount of emissions that each state receives from other states;

3. A state sulfur dioxide emissions budget that is to be allocated across existing sources or that can be sold on a Federal Emissions Options Market;

4. An emissions tax on all emissions that is used to offset the cost of regulatory control and partially to provide corrective action to mitigate acid rain damage;

5. A State Implementation Plan (SIP) process that decides and reports how states have allocated their emissions budget and demonstrates that they are in compliance with the Clean Air Act;

6. The awarding of state reduction incentive credits to states that reduce their emissions below the initial emissions budget allocation. These credits can be exchanged on the Federal Emissions Options Market.

The first provision is justified on the grounds that all states should make an effort to reduce sulfur dioxide emissions because they present environmental and health risks that are of national concern. Taking the 1980 emissions reported in Table 8.1 as a starting point, each state would be required to reduce its total emissions by a fixed precent, say by 38%--the amount that would be required to achieve the 10 million ton reduction proposed in H.R. 3400 and in other congressional proposals.

As a matter of equity based on principles of state sovereignty, no state should be the dumping grounds for another state's environmental pollutants. Therefore, a maximum level of sulfur dioxide emission imports into a state is permitted. Reliable estimates of the migration of emissions between states exist. Table 8.4 presents the transport of emissions among the southern states showing each state's sulfur dioxide emissions, the emissions received from other states, the amounts received from the midwestern states, and the proportion of the state's emission that falls within its borders.

While the exact level of allowable imports is difficult to set, several guidelines can be relied on. For example, no state should involuntarily import more emissions from other states than it creates itself. This principle provides an extra incentive for states to reduce their own generation of sulfur dioxide because they will, at the same time, be reducing the amount of allowable imports. Based on Table 8.4, both Arkansas and Virginia will qualify for immediate relief.

A second guideline for establishing the maximum imports allowed into a state is that if a state can demonstrate persistent and permanent environmental harm to its ecological system as a result of the continuation of present levels of sulfur dioxide, then the Federal Environmental Protection Agency must purchase sufficient and appropriate emission permits so that necessary emissions reduction is achieved.

The third provision of this proposal is the assignment to each state of a sulfur dioxide emissions budget that is less than the maximum amount of state emissions. Each state's allocation of the national emissions budget should be made on the basis of a scientific and technical consensus about what type of polluting activities are most easily modified to reduce emissions. In order to provide incentives for emissions reduction each state's emissions budget must be below the maximum allowable emission level. States, therefore, must purchase emission permits beyond that budget allocation, up to the maximum level, if they require additional emissions. The original budget emissions can either be allocated across existing sources within the state or the emission permits can be sold to other states on the Federal Emissions Options Market.

Another provision of this proposal that must be considered by policymakers is the amount of the emissions tax. This tax should be set at least at the level that will fund the cost of the regulatory program and reimburse federal, state, and local efforts to mitigate acid rain damage. This method of setting the emissions tax is an "actual cost approach" that is much simpler than attempting to estimate the social costs of acid rain damage.

The implementation mechanism constitutes the fifth requirement of this proposal. The existing State Implementation Plan provision of the Clean Air Act can be relied on to administer the allocation of the emission budget, to collect the emission tax, and to

make state decisions about entering the Federal
Emissions Options Market. SIPs must continue to
satisfy the current detailed reporting requirements to
demonstrate that they have complied with the Clean Air
Act.
     Finally, an additional incentive for state
emission reduction is included in the proposal through
the awarding of reduction incentive credits to states
that reduce emissions below their initial budget
allocation. These credits can be exchanged on the
Federal Emissions Options Market, cashed in to the
United States Treasury, or be used to offset other
emissions that exceed the state's emissions allocation.
This provision provides a double incentive for
reduction. First, a state can sell unused emission
permits from its initial allocation of the emission
budget, and then, second, it will receive an additional
incentive through the award of a reduction credit.
     This state-based reduction approach offers several
provisions that differ from current proposals such as
H.R. 3400. First, this proposal is, obviously, state-
based and does not automatically limit the "top 50
emitters" as do many of the current proposals. Second,
this proposal affects all emitters of sulfur dioxide by
virtue of an emissions tax. This incentive encourages
small emitters to make the same effort as larger
producers and the tax can be passed on to the ultimate
consumer of the good or service that causes the
production of sulfur dioxide in the first place.
Therefore, the social costs of sulfur dioxide emissions
are being internalized into the normal price of the
relevant commodity. In the case of electricity, the
emissions tax can be passed on through the pricing
regulations of state utility commissions to be borne by
the ultimate consumer of electricity regardless of
location. Third, a mimimum amount of emissions relief
or protection is provided to each state, "as a matter
of right," by establishing a maximum state emission
import that any state can receive. If this maximum is
exceeded then the Federal Environmental Protection
Agency must enter the Federal Emissions Options Market
and purchase emission rights in a way that will reduce
the state's emissions imports.
     A fourth attraction of this proposal is that it
establishes a Federal Emissions Options Market that
provides a forum for the exchange of emission rights.
Like commodities or monetary markets, this market will
generate pricing information that reflects the value of
emissions rights. Like other open markets participants
will include a wide variety of consumers, producers,
governmental institutions, and citizens concerned about
the negative consequences of sulfur dioxide emissions.
     Fifth, the current State Implementation Plan

provision of the Clean Air Act is relied on to allocate the state's emissions budget and to buy or sell emissions rights as the state sees fit.

Sixth, reliance on an economic market for solving an environmental problem forces the states to establish environmental and economic development priorities that can result in better coordinated policy. For example, efforts to improve states' economic conditions through various reindustrialization programs must now consider the environmental aspects of certain types of economic-environmental activity.

Seventh, the costs of sulfur dioxide emissions can be passed on to the consumer or the final user of the commodity responsible for the generation of the emission. Therefore, "innocent" parties are not being asked to bear the financial burden of emissions reductions.

Eighth, where several aspects of this proposal promote administrative and managerial flexibility, this proposal is attractive because it can be easily modified over time by changing the maximum levels of emissions allowed, or the emissions tax rate, or the amount of allowable imports. Thus, once the regulatory mechanism is in place formal modification is not difficult. Alternatively, local, state, and federal governments as well as private citizens can have the same effect as formal modification by strategically purchasing the emissions rights form the appropriate emitter or state.

Ninth, and finally, this proposal focusing on sulfur emission reduction can be easily dovetailed with other emission reduction programs for nitrogen oxides, hydrocarbons, particulates, etc. by establishing "conversion offsets" or "tradeoff credits" that can be earned by the reduction of particular emissions and used to purchase additional emissions for another pollutant.

SUMMARY

The regulatory problems surrounding acid rain are many. Among them are included a diversity in regional sulfur dioxide emissions, environmental harm, economic interests, and environmental concern. Traditional regulatory schemes based on uniform standards are both politically and economically unacceptable in addressing the acid rain problem. A state-based regulatory approach relying heavily on economic incentives offers numerous advantages compared to the current proposals now before Congress. The proposal presented here, while requiring the creation of a Federal Emissions Options Market, incorporates the existing regulatory

structure currently enforcing the Clean Air Act.
The six provisions developed above constitute a
regulatory approach that is practical and manageable.
While several political decision-makers are involved
and are required to make difficult decisions for this
proposal to be implemented, the scientific and
technical information exists that makes this approach
workable. While increasing the administrative and
managerial flexibility of emissions reduction, compared
to the current "command and control" approach, this
proposal gives policymakers discretion in setting
ceilings on state emissions, the level of allowable
emission imports, and the amount of the emissions tax.
Like all policy decisions, the ones required in this
proposal will undoubtedly generate political
controversy and conflict in their design and
implementation.

REFERENCES

Ackerman, Bruce A., and William T. Hassler. **Clean Coal/Dirty Air**, New Haven: Yale University Press, 1981.

Anderson, Fredrick, et al. **Environmental Improvement Through Environmental Incentives**. Baltimore, Md.: The Johns Hopkins University Press, 1977.

Baumol, William J., and Wallace E. Oates. **Economics, Environmental Policy, and The Quality of Life**. Englewood Cliffs, N.J.: Prentice-Hall, 1979.

Blackman, Sue Anne Batey, and William J. Baumol. "Modified Fiscal Incentives in Environmental Policy." **Land Economics**. 56 (November 1980). 417-431.

Congressional Research Service. "Acid Rain: Current Issues" Issue Brief Number IB83016, Washington, D.C.: CRS, 1984.

The Conservation Foundation. **State of the Environment: An Assessment at Mid-Decade**. Washington D.C.: The Conservation Foundation, 1984.

Cowling, E.B. "A Status Report on Acid Precipitation and Its Biological Consequences as of April 1981" In Frank M. D'itri, ed. **Acid Precipitation: Effects on Ecological Systems**. Ann Arbor, MI: Ann Arbor Science, 1982, 2-20.

Crandall, Robert W. **Controlling Industrial Pollution: The Economics and Politics of Clean Air**. Washington, D.C.: The Brookings Institution, 1983.

Edison Electric Institute. "Evaluation of H.R. 3400-- The 'Sikorski/Waxman' Bill For Acid Rain Abatement." Washington, D.C.: Edison Electric Institute, 1983.

Gallogly, Margaret R. "Acid Precipitation: Can The Clean Air Act Handle It?" **Boston College Environmental Affairs Law Review**, 9 (1980-81). 687-744.

General Accounting Office. **A Market Approach to Air Pollution Control Could Reduce Compliance Cost Without Jeopardizing Clean Air Goals.** [PAD-82-15]. Washington, D.C.: General Accounting Office, 1982.

Ingram, Helen M., and Dean E. Mann. "Environmental Policy From Innovation to Implementation." In Theodore J. Lowi and Alan Stone eds., **Nationalizing Government: Public Policies in America.** Beverly Hills: Sage Publications, 1978. 131-162.

Kelman, Steven J. "Economic Incentives and Environmental Policy: Politics, Ideology, and Philosophy." In Thomas C. Schelling ed., **Incentives for Environmental Protection.** Cambridge, Mass.: The MIT Press, 1983. 291-332.

Levin, Michael H. "Getting There: Implementing the 'Bubble' Policy." In Eugene Bardach and Robert A. Kagan, eds. **Social Regulation: Strategies for Reform.** San Francisco: Institute for Contemporary Studies, 1982. 59-92.

National Clean Air Coalition. "Acid Rain in the South." Washington D.C.: National Clean Air Coalition, 1984.

National Research Council, National Academy of Sciences. **Acid Deposition-Atmospheric Processes in Eastern North America.** Washington, D.C.: National Academy Press, 1983.

National Wildlife Federation. "Weathering the Storm." **National Wildlife**, 1983.

Office of Technology Assessment. **Acid Rain and Transported Air Pollutants: Implications for Public Policy.** [OTA-O-204]. Washington, D.C.: U.S. Congress, OTA, 1984.

Peterson, Jerold. "Estimating an Effluent Charge: The Reserve Mining Case." **Land Economics.** 53 (August 1977). 328-341.

Repetto, Robert. "Air Quality Under the Clean Air Act." In Thomas C. Schelling ed., **Incentives for Environmental Protection.** Cambridge, Mass: The MIT Press, 1983. 221-290.

Rhodes, Steven L. "Superfunding Acid Rain Controls: Who Will Bear the Costs?" **Environment.** 26 (1984). 25-32.

Part III

International-Comparative Dimensions

The acid rain phenomenon has been a source of persistent international tension and continuing diplomatic negotiations among a number a neighboring nations in Europe and North America. In the North American continent, the failure of bilateral efforts begun in 1979 to settle the problem of transboundary air pollution between the United States and Canada has injected a strain of bitterness and resentment in U.S.-Canadian relations which have otherwise been noted for their harmony and close partnership on a number of fronts. Within Europe, discernible movement to ameliorate transboundary acid rain problems has been evident on the diplomatic front at least since 1982.

In Europe, the problem of acid rain is highly visible and environmental damage attributed to acid deposition is apparently widespread. The acidification of lakes and surface waters in Scandinavian countries, and the devastation of forests in West Germany have been widely highlighted in the media, as well as in scientific and other reports. Despite the notable differences between the United States and Western Europe with respect to the distribution of emissions and deposition, the nature and extent of damage to lakes and forests in continental Europe and Scandinavia are viewed by some as a portent of the future for surface waters and forests in the United States.

Canada and some West European nations have initiated plans to address the problem of acid deposition in a cooperative fashion. Canada has recently announced plans to reduce SO_2 and NO_x emissions from smelters and automobiles by 50 percent in its eastern provinces over the next nine years. In Europe, several nations (e.g., the Federal Republic of Germany, France, Norway, Sweden, and Great Britain) have joined with Canada to form the "30 Percent Club"--a multinational effort to address the acid rain by lowering SO_2 and NO_x emissions by 30 % over a number of years.

The following chapters in this part by Ernest J.

Yanarella and John E. Carroll provide different perspectives on international efforts to address the acid rain problem, and seek to define some of the major policy and political implications for the United States flowing from those international developments.

Yanarella's chapter presents a critical comparative analysis of the "promise and limitations" of American environmentalism by juxtaposing the structure, orientation, and philosophical/cultural foundations of American environmental groups with that of the West German Green Party. The philosophical basis for this comparison is the distinction between "environmentalism" and "ecology"--terms which he suggests are not synonymous, but are rather "two fundamentally different ways of looking at, and relating to, nature." With its roots in the American preservationist movement and progressive conservationism, American environmentalism is often "fixated on narrow technical issues of resource conservation and pollution reduction" pursued in terms of the narrowly instrumentalist terms of interest-group politics. In contrast, he, argues, the Green Party is a "political force cultivating a changed relationship between human beings and nature."

The structure and orientation of the West German Green Party thus stands in marked contrast with American environmentalism. While American environmentalists tend to embrace the precepts and practice of interest-group liberalism, the Green Party is "striving to infuse their political thought and practice with a new cultural paradigm." Yanarella analyzes Green politics in those terms, and suggests that the Greens offer "some important clues and guides for the further development of environmentalism in the United States beyond the constraints of interest-group politics and legalistic-moralistic prescriptions." Although there are structural and other differences between West German and American political systems which limit the extent to which one can draw upon the Green experience, he sketches some of the "germane lessons" for a "political ecological framework" applicable to environmental politics in America and the acid rain debate.

John Carroll views the acid rain debate in terms of the relationship between the "political geography" of acid rain and international diplomatic initiatives taken to address the problem. He argues that the extent to which a nation's land and water resources are vulnerable to damage by acid rain plays an important role in the "diplomatic behavior of that nation relative to acid rain and international relations." Given the experience in Europe and North American with acid rain, it is clear that it has become a serious problem for "international diplomacy." Carroll then reviews the European experience with acid deposition and the various mulilateral efforts to address the problem with the conviction that some guidelines can be defined which will be useful in "developing international arrangements and achieving diplomatic solutions to the acid rain issue in North America." From this survey, he concludes that the approach of the "30 Percent Club" represents an intriguing model for repairing relations between the United States and Canada over the acid rain cross-boundary issue.

Because the acid rain problem has been responsible for some very serious diplomatic tensions between the United States and Canada in recent years, Carroll believes that a resolution of this issue is a test of political will, and can be approached in an incremental fashion which acknowledges the partial truths held by competing political alliances involved in the debate. A "reasonable compromise" can be forged which will represent at least a first step toward a more complete resolution when such favorable circumstances crystallize in the future. According to Carroll, that compromise plan is a variation on the 30% solution. He thus urges the United States to agree to reduce its emissions of SO_2 by 30% over a ten-year period. And this goal, he contends, can be accomplished without the use of scrubbers. Instead, he argues, reductions could be achieved by requiring coal washing, utility least-emission dispatching, and the use of low-sulfur coal. On the last feature, he concedes that the use of low-sulfur coal should not be so large as to threaten "the viability of the high-sulfur coal industry."

As the chapters in this part clearly indicate, the problem of acid rain transcends national boundaries and poses significant challenges for the development of multinational institutional arrangements for resolving difficult environmental problems such as acid rain. In addition, the authors in this part suggest that an understanding of political developments related to the handling of the issue of acid rain within particular national arenas in Europe and Canada can offer important lessons to policy "insiders" and "outsiders" alike. On the one hand, it can offer policymakers in this country some perspective on the limits of the politically possible while, on the other, it can present to American environmentalists some insight into ways of restructuring those prevailing constraints on the politically possible.

9
Environmental vs. Ecological Perspectives on Acid Rain: The American Environmental Movement and the West German Green Party

Ernest J. Yanarella

INTRODUCTION

In spite of the heavy involvement of environmental groups like the National Clean Air Coalition, the Natural Resources Defense Council, and the National Wildlife Federation in the policy struggle for acid rain control, the results to date have been disappointing. Without discounting the prodigious political and ideological obstacles which have impeded their long-term campaign to get enacted a mitigation program in the United States, it may be time to consider the possibility that part of the problem may lie in the organization and strategy of this key element of the pro-control alliance. In recent years, American environmentalism has been subjected to a fundamental critique from the right.[1] While some of the specific indictments lodged against present-day environmental and natural resource policies carry a measure of truth, the right-wing critique is founded upon romantic and idealistic images of the capitalist economy and of marketplace practices which were long ago transcended by internal transformations within the capitalist system itself in the United States and most other advanced capitalist societies.[2]

A more fruitful and politically progressive perspective from which to render critical judgment on the promise and limitations of environmentalism lies in contrasting this concern for nature with an alternative manifestation--ecology. Although environmentalism and ecology are often viewed as synonymous, they actually represent two fundamentally different ways of looking at, and relating to, nature. As Leo Marx has pointed out, whereas environmentalist thought has historically adopted the perspective that "nature exists apart from, and for the benefit of, mankind," the ecological view sees humankind and its humanly-made (or secondary) environment as "wholly and ineluctably embedded in the tissue of the natural process."[3] Moreover, environmentalism as a movement has tended to be a loose coalition of nature lovers, hikers, and campers represented overwhelmingly by white, middle-class professionals. Ecology, on the other hand, has emerged in recent years as a political force cultivating a changed relationship between human beings and nature, calling for the shaping of humanly-scaled technologies compatible with the natural patterns and forms of Mother Earth, and struggling to build the foundations of human community based upon social justice and ecological norms.[4] Opting for one vision or the other entails, at deepest levels, alternative political loyalties and strategies, since the basic thrust of environmentalism is utilitarian and reformist, while ecology--because it goes literally and figuratively to the roots--is reconstructive and radical.

This philosophical point may be made more concrete by comparing organizational forms and political strategies within American environmentalism with those of the West German Green Party. It will be the burden of this chapter to show that the structure and orientation of the latter is a more promising route for responding to policy stalemate in the environmental realm in late capitalist societies and, moreover, that it offers some important clues and guides for the further development of environmentalism in the United States beyond the constraints of interest-group politics and legalistic-moralistic prescriptions.

AMERICAN ENVIRONMENTALISM; FROM MOVEMENT TO
TO PROFESSIONAL INTEREST GROUP

Contemporary environmentalism in America owes its impetus to the conservation and preservation movements, which emerged between 1890 and 1920. Responding to the reckless pillage and wanton destruction of the nation's natural resources during the preceding decades, "progressive conservationism" took root and flourished out of an alliance of state agencies and "responsible" corporations which recognized the need to promote careful resource management through centralized administration, technical efficiency, and long-range planning.[5] Far from being economically progressive, though, this initial phase of conservationism proved not to be "an attempt to control private, corporate wealth for public ends," but instead was an incipient expression of the emerging foundatons of the corporate state.[6] Elitist in outlook and directed by engineers and other specialists, it conceived of stewardship of nature's treasures in essentially technical terms and administrative criteria. Thus, the proneness of some contemporary environmental groups (such as the Izaak Walton League) and environmental protection agencies (both state and federal) to translate political and economic problems into technical-administrative issues can be explained in part by the way

conservationist ideology was nurtured on the "gospel of efficiency" in the early 1900s.
The Sierra Club and the National Audubon Society, on the other hand, are representatives of the early preservationist movement. In contrast to the conservationist stream of modern-day environmentalism, preservationist organizations from the beginning have been most concerned with habitat, not sustenance, that is, with the value of setting aside wilderness areas and parklands for the enjoyment of present and future generations.[7] Institutionally, preservationist objectives coincided nicely with the bureaucratic self-interest of the Department of Interior in enlarging its mission and responsibilities while, in class terms, such aims have comported with the leisure activities of middle- and upper middle-class professionals. It is not surprising then that the board of directors of many of these contemporary preservationist societies "reads like a directory of corporate America."[8] Only in the past decade or so have these organizations come into with America's mega-corporations, as the real and contrived dimensions of the energy crisis and looming resource scarcities have triggered intensive corporate campaigns to exploit indigenous energy and other natural resources locked up in federally-controlled wildlife and wilderness preserves.
The political and organizational legacy of these earlier voluntary associations has weighed heavily upon environmentalism in policy debates since the late sixties. To a considerable extent, the dominant concerns of the modern environmental movement may be seen as "integrating the habitat and sustenance concerns of the efficiency and preservation movements."[9] Furthermore, the heritage of the past has also acted to enhance the preeminent position of these elitist, hierarchical organizations within American environmentalism and to shape it politically into a largely middle-class, reformist movement guided by status concerns taking the form of symbolic moral protest. On the other hand, latter-day environmentalism differs from the voluntary associations of the past in a number of key respects. For one thing, most mainstream environmental groups are now backed by professional staffs incorporating politically-savvy lobbyists, environmental lawyers, and impact scientists who have shown great skill at holding their own against their corporate counterparts in legislative and judicial combat.[10] On the positive side, the size, budgets, and capabilities of these organizations mean that professional environmentalism is a powerful force that must be reckoned with. On the negative side, however, their hierarchical structure, mass-mailing appeals, and largely passive memberships tend to contribute to a form of alienated politics which is not really overcome by periodic pseudo-participatory waves of

mobilization to write letters, send telegrams or make phone calls to legislators to protest environmental scandals or to support key bills.[11]

A second feature of the new environmental movement which distinguishes it sharply from its historical progenitors is its tremendous scope and variation. Because environmentalism is not a static force of unchanging ideas and concepts but a living creation of men and women inhabiting a changing social and natural world, it has given rise to groups like the "Friends of the Earth" and the multiplicity of anti-nuclear groups and taken more democratic, decentralized organizational expressions.[12] Simultaneously, it has also spawned specialized professional groups like the Natural Resources Defense Council and the Environmental Defense Fund which have been principally concerned with litigating environmental issues or lobbying on behalf of environmental legislation in the Congress. Again, on the positive side, this broad conglomeration has encouraged an informal division of labor among increasingly specialized groups dotting the environmental scene whose power can be consolidated by the forging of coalitions centered around the principal articulator of each narrow issue.[13] But one of the major problems with the character of environmentalist involvement in the political dispute over acid rain controls is that the more novel and holistic perspectives of the former organizations have been overshadowed by the predominance of the strategies and tactics of the latter in acid rain politics. One troubling consequence of legalistic/lobbying methods is that the kind of interest-group syndicalism that it promotes erodes public confidence in politics as a means of settling disputes and of fashioning sound policy, and thus fosters political cynicism and system delegitimation."[14]

No less serious, this obeisance to interest-group politics has almost invariably led environmentalists to depreciate working-class and labor interests in energy-environmental disputes. Having succeeded in achieving legitimacy for environmental considerations in policy-making arenas, environmentalists often adopt a highly instrumentalist approach to policy disputes on issues like air pollution and acid rain--thus, alienating organized labor, the poor, and blacks who then form a "popular front for accumulation" with corporate elites and right-wing politicians.[15] Yet real growth in the social constituency and the political clout of the environmental movement, as Allan Schnaiberg reminds us, surely requires at a minimum a multi-class strategy "of seeking benefits for social groupings that do not constitute the main membership --or even the most supportive public--of the movement, through the ongoing influence of the movement."[16] In this shift from a narrow environmentalist to a socio-ecological perspective, a key objective would be "to permit enhancement of future

mobilization potential [across classes], to consolidate
and extend social environmental gains".[17] In the absence
of incorporation of labor and equity considerations into
the environmentalist outlook and agenda, the movement is
likely to remain fixated on narrow technical issues of
resource conservation and pollution reduction and thus
remain an easy target for further elite cooptation.
    American environmentalism thus stands at a fateful
crossroads. On the one hand, it is weighed down by its
anchorage in middle-class moralism and the culture of
professionalism; on the other, it is prodded by a growing
sensitivity to earthly limits and to novel social
alternatives and possiblities visible on the cultural
horizon. When environmentalists look back and are
animated by the originating values and ultra-pragmatism
of the early voluntary associations, they shield
themselves from what they do not wish to recognize--
namely,"how their personal, professional, and
organizational acts may contribute to a strengthening and
legitimizing of precisely those processes and
institutions that are destroying the natural world, their
community, their health, their progeny, or, most
important, their chance to empower themselves in a
different kind of system that exploits neither nature nor
people".[18] When they look forward, they see the vague
outline of an alternative society constituted on a
different set of relations between human beings and with
nature, a new science and technology infused with
emancipatory values, and new forms of social
organization in the workplace and the larger community
governed by truly democratic norms and procedures and
constructed according to human scale.

THE WEST GERMAN GREEN PARTY: HARBINGER
OF POLITICAL ECOLOGY

    Like the contemporary environmental movement in the
United States, the Green Party in the Federal Republic of
Germany is a creature of its political culture and its
times. Germans have long felt a strong attraction to
nature, particularly their forests. The poets Goethe and
Hesse rhapsodized on the spiritual character of the trees;
the composer Beethoven and the philosopher Heidegger drew
creative inspiration from the German forests; and
millions of Germans today take to the woods each month to
overcome the modern division between human beings and
nature.[19] During the Nazi period, the German romantic
tradition took a dark detour when Hitler appropriated the
themes of nature and ecology on behalf of regressive
political urges.[20] But, beyond this enduring cultural
attachment to German landscape, the emergence of the
Greens around 1980 was greatly affected by the political

trajectory of the Social Democratic Party (SPD) after the Second World War from a party of "fundamental opposition" to a party of "responsible opposition" to a "catch-all" party-in-government of the political establishment. For, as the SPD cashed in its socialist democratic vision for broader constituencies and political votes it helped to create a situation by the sixties and seventies in West Germany where generational conflicts and changing political perceptions about the nation and the world created an assortment of new needs and interests among unrepresented groups--particularly the post-war generation of youth--which SPD leaders could neither aggregate nor coopt.²¹

Triggered partly by the SPD's attempt to channel new political currents into the party, partly by spontaneous efforts at the grassroots, a wave of extraparliamentary opposition groups burst forth on the German scence in the mid-seventies and took the form of citizen initiatives (**Burgeriniativen**). Organized around anti-nuclear, peace, feminist, and other single-issue foci, these citizen action groups came to number an estimated 38,000 with a membership of between two and three million.²² For a time, they operated largely at the local level, employing mobilization strategies to promote citizen education and participation in issues which the established parties failed to address or responded to only poorly. Unified ideologically by a nascent ecological view of the political world, many of these groups increasingly turned their attention to the need for greater coordination and to the merits of parliamentary involvement when different strategies of anti-nuclear protest failed to halt West German development of nuclear power. While nationwide coordination of the ecology movement through the Federation of Citizen Initiatives for Environmental Protection (BBU) proved to be a valuable interim step toward organizing anti-nuclear and other forms of political protest, the feeling became widespread among key figures associated with the BBU that the abolition of nuclear power and the achievement of disarmament and sexual equality could only occur through fundamental changes in German society.²³ Consequently, beginning in 1977, local environmental party lists of candidates were placed on the ballots which then proliferated throughout West Germany under the label, the Green List Environmental Protection (GLU). Out of the development of these local lists by the German eco-left and the involvement of two splinter parties of varying ideological orientations (the AUD and the GAZ) was forged the parliamentary arm of extra-parliamentary opposition in the form of the Green and Alternative List Parties (G/AL).

From its formal institution in January 1980 to the federal elections in March 1983, the Green Party has grown from a local and regional presence on the German political landscape to a party of fundamental opposition within the German Bundestag with a **Fraktion** of 27 parlimentary representatives.[24] Thusfar rejecting inducements from the traditional parties to act responsibly and play by the parliamentary rules of the game, the Greens have raised profound questions about the conventional rules of the game and have further put in doubt the efficacy of the traditional party. Conceiving itself as half-party/half-movement, the Green Party has regarded its parliamentary involvement as ancillary to its primary role of being the political voice of the extra-parliamentary opposition housed in the many citizens' initiatives forming its base (**Basis**). As a result, the party has employed its access to the parliamentary system mainly to provide information support to these grassroots movements, although it has also used its parliamentary position to inject principled policy alternatives into the legislative process.

The key inspiration of Green politics in West Germany lies in pervasive awareness at its political base and among its leadership that, not only Germany, but the entire world community is in the throes of a multi-dimensional crisis affecting every facet of our personal and collective lives. Rather than responding to this unprecedented situation with ideas and programs borrowed from traditional ideologies, the Greens are striving to infuse their political thought and practice with a new cultural paradigm.[25] The old cultural paradigm-- which formed the seedbed of Western thought and action from the early modern epoch of the sixteenth to eighteenth centuries and gave impetus to the Enlightenment philosophy and industrial capitalism--was characterized by an ontological outlook which divided the world into opposing dualisms (self/others, mind/body, human-kind/nature, knower/known , male/female) and sanctioned the dominance of one pole or sphere over the other. In the process, this modern philosophical worldview contributed to the unleashing a series of dynamic cultural and political-economic processes which fostered the degradation of nature and the disintegration of human community.[26] Unlike environmentalism which remains stuck fast in the dualisms of the modern philosophical paradigm, Green politics is groping towards a fundamental reconceptualization of the world which restores the existential and intentional ties connecting the realms ruptured by the early modern revolutions in science, religion, and economics in the West--or at least to sees those relations (e.g., humankind/nature and male/female) in their identity and nonidentity.[27]

More concretely, the political platform of the Green Party is supported by four key pillars: ecology, social welfare, grassroots democracy, and nonviolence.[28] Removing man from the center of the universe in favor of a cultural and political framework which relocates human beings within a more encompassing system of relations, this program aims at preserving valued relationships and putting a brake on uncontrolled industrial growth and social "progress" through an "attempt to avoid irreparable losses and rreversible catastophes in the areas of peace, the natural environment, and civil and human rights."[29] This has led Green Party representatives to adopt a set of policy recommendations and proposals which undertakes a careful ecological cost-accounting of deployment of U.S. cruise and Pershing II missiles on their nation's soil, continued industrial pollution of the air and water, predominance of high-technology, allopathic medicine, and retention of the **Berufsverbot** laws banning suspected radicals from public employment.[30] In other words, Green politics has been motivated and directed by a holistic, ecological critique of late capitalist society--including its politics, culture, economy, and social relations.

In the realm of "environmental" policy, the Greens have taken the lead in mitigating the damage incurred within West Germany and across its boundaries by the phenomenon of acid rain. The policy reversal on the need for international measures to alleviate transboundary acid rain damage in Western Europe made by the SPD-led West German government at the Stockholm Conference on the Acidification of the Environment in 1982, as well as the reluctant response to palpable evidence of the dying of over a third of German forests taken by the Christian Democratic Party (CDU) since its rise to coalition power in 1983, can be attributed in large measure to the ceaseless mobilization campaign of the Green Party against **Sauren Regen** (acid rain) and **Waldsterben** (the dying of the forests).[31]

In contrast to the most active environmental groups working on the acid rain problem in the United States, the Green Party has recognized this environmental issue as being inextricably connected to economic policy and employment policy. Thus, although The Greens have been active in proposing measures to mitigate the effects of acid rain through technological means (e.g., by introducing into the Bundestag in May 1983 a multi-stage proposal for reducing sulfur dioxide emissions levels by the year 2000 to 1.9 percent of the levels existent in 1983 and by calling for a 100 k.p.h. speed limit on major German highways and the use of unleaded gasoline by German automobiles), they have not neglected to place the problems associated with combatting industrial pollution and the other costs of pro-growth policies into a policy matrix which links their ecological and controlled growth programs to plans for stimulating full employment.[32] As several sympathetic students of Green politics have acknowledged, the Green economic and employment program remains at an early stage of development and refinement. But, insofar it has been outlined, the program has been organized around the effort to take seriously their democratic, decentralist vision of an alternative society by devising mechanisms for using public capital for social investment in environmentally benign, alternative technologies and worker-controlled, more labor-intensive industries and by creating programs for easing the necessary adjustments to a new industrial foundation for a steady-state society.[33]

ACID RAIN IN POLITICAL ECOLOGICAL PERSPECTIVE

The structural differences between the German and American political systems frustrate any aspiration to draw easy lessons from the Green Party experience for American environmentalism. As the preceding section has shown, the Green Party was able to take advantage of the German parliamentary system with its proportional representation. In addition, the Greens have benefited enormously from a number of features distinctive to West Germany--specifically, the fact that their country is "a densely populated, heavily industrialized nation where the limits to growth are visible at every turn, where the madness of nuclear deterrence has made them prime candidates for thermonuclear holocaust, and where the level of affluence allows 'big picture' reflection."[34] By contrast, environmentalists in the United States must contend with a huge, sprawling land whose enormous natural wealth and lush forested regions seem to belie the destructive impact of industrial production, whose sheer size allows its military-industrial complex and strategic storehouses to fade into the natural landscape, and whose individualistic, possessivistic culture has engendered an American Dream of social mobility and material affluence which immunizes itself from criticism by blaming the economic system's victims for failing to win the race. Still, as suggested above, germane lessons can be found in the Green Party's more holistic outlook and its attentiveness to labor and social equity issues.

Beyond these cues, a political ecological framework also yields a number of observations pertinent to conceptualizing and mitigating the acid rain problem:

--**A synoptic view of the acid rain problem is essential to its solution**: A holistic view of the causes and ramifications of acid deposition is necessary to confront it as an issue for creative public policy. Unfortunately, such an orientation is precisely what most environmental groups and individual scientific subcommunities eschew. Hyperspecialization, narrowly focused research, and a segmented conception of the field of study all contribute to a professional incapacity to, or a trained ignorance of, the dynamic interaction between and among natural ecosystems and social systems. But, from the ecological perspective, if everything is connected to everything else, the problem of acid rain must not be understood in isolation from other closely-related problems of environment and ecology.

Two things are implied by this statement. First, acid deposition is not a discrete problem which will yield to a single, even comprehensive, policy. As Robert Boyle and his son put it, "it is important to bear in mind that acid precipitation is only one manifestation, albeit a very serious one, of the widespread pollution of the atmosphere."[35] Secondly, given the enormous complexity of the interactions among individual natural ecosystems and social systems constituting our overall Ecosystem we call the planet Earth, it is probably impossible in principle to expect scientists to derive definitive answers to the many puzzling questons about the behavior of acid deposition in its various forms. Consequently, political intervention must always operate in a context of scientific, technical, and other uncertainties. Naturally, there is no guarantee that abatement efforts will succeed but, as the recent OSTP interim report notes, "recommendations based upon imperfect data run the risk of being in error; recommendations for inaction pending the collection of all desirable data entail even greater risk of damage."[36]

--**If the Earth is a self-regulating organism, we may not yet be in a crisis situation on acid rain:** One obvious problem in the acid rain debate in the United States is that environmentalists and industrialists share a common view of the earth as a mechanical universe or mathematical matrix, potentially comprehensible through a formal quantitative model. But what happens if the earth is a complex, living organism on the order of James Lovelock's **Gaia?**[37] On this view, the biosphere is interpreted as "a self-regulating entity with the capacity to keep our planet healthy by controlling the chemical and physical environment"[38]--and, it might be added, even by adjusting within limits to large-scale human interventions and foul-ups.

To place the acid rain problem within this conception of the biosphere is not however to adopt a complacent attitude to the control or abatement of its negative consequences. An ecological view of the Earth and its human inhabitants implies that Nature is not a mere passive object either for rapacious exploitation in the name of economic growth or for upper middle-class adulation in the name of environmental regulation or conservation. It is much more like an organic, though nonsentient, body with a pulse and rhythm and immune system very much akin to the medieval identification of Nature with a woman's body.[39] So conceived, the biosphere is not a mechanistic or inert substrate passively responding to human industry and intervention. Indeed, some of the climactic oddities of the past five to ten years may be evidence of our Earthly mother attempting to adjust to the growing acts of ecological violence perpetrated against her body. Thus, in a most peculiar way supportive of the pro-industry arguments, it may be that we are not yet in a crisis state with respect to the impact of acid deposition upon aquatic, forestry, and other ecosystems.

This concession does not mean that policy actions aimed at meeting the evolving dangers to the environment and to human health are premature or should await definitive answers from science. For, on the one hand, Western science is conservative, skeptical, and consensual in its approach and at least in terms of the dominant philosophy of science underpinning the self-understanding of scientific practice by the community of scientists, it can never reach absolute verities. This would seem especially true for an organism as enormous and as complicated as the Earth in its continuing interaction with its human residents. On the other hand, even a self-adjusting organism on the scale of the biosphere does have points of strategic vulnerability whose destruction may irreversibly damage the health and functioning of the Ecosystem as a whole.[40] The OSTP report, for instance, has openly worried about the impact of acid rain upon the capacity of microorganisms to recycle nitrogen and carbon in the food chain.[41] Such an injury to the biosphere would pose grave consequences for nature and humankind, and conservative, cautious science would probably recognize this mortal blow far too late for human intervention to correct it.

--**Other nations are adopting more responsible policies on acid rain than the United States**: Clearly, other countries--particularly those in Scandinavia and several in continental Europe--are doing a much better job than the United States in confronting acid rain as a domestic and as a transboundary problem.[42] Part of the reason for their greater sensitivity to this national and trans-national anomaly relates to the wider ideological spectrum informing their party systems, the great strength and importance of certain forestry and fishing industries to their economies, and the more deeply ecological consciousness influencing public attitudes and shaping the vector of political forces operating in their respective political systems.

It must be remembered that one reason for the gaps in scientific knowledge about acid deposition in the United States has to do with the late institutionalization of environmental consciousness in this country and the relative weakness of environmental (public) versus commercial (private) interests and values in the United States in comparison with certain West European and especially Scandinavian countries. Many scientists go where the research funds are and the history of acid rain research in America is a catalogue of lax, intermittent interest at the federal level in funding research and halting, interrupted investigative efforts within the scientific community to clarify its nature.[43]

Energy-environmental quandries like acid rain may force scientists and environmentalists to consider anew an alternative post-modern/ecological orientation: The release of an EPA-sponsored report conceding serious and inevitable climactic changes globally due to the "greenhouse effect," as well as the publication of a continuing profusion of studies by scientists on the growing threats posed by acid deposition and other ecological hazards (ozone, heavy metals, etc.) place before the scientific community a basic challenge to the modernistic moorings of its technicist orientation and objectivist philosophy of science.[44] Indeed, one is almost tempted to use Hegel's notion of the working out of the cunning of Reason to explain the possible transformation from a modern to a post-modern philosophical framework--metaphorically because there are no guarantees in philosophy, religion, economics or politics that it will occur at all or in time.

To elaborate, it is startling how enmeshed is the early history of the mining of coal and metals--the twin culprits in the acid rain phenomenon--in the overthrow of medieval norms and sanctions in Western culture and economy and in the crystallization of modern assumptions and permissions--especially toward an alienated view of nature as wholly other and as an object of domination for human gratification. In different ways, Carolyn Merchant in her book, **The Death of Nature**, and Lewis Mumford in his classic work, **Technics and Civilization**, show how, in the process of the breakdown of medieval normative constraints against mining of the earth, the experience and environment of underground mining of coal and metal ores reflected key features of the conceptual universe of the modern philosophical outlook informing the vision of pioneering philosophers like Bacon and Descartes and the work of early modern physicists.[45] "The mine," argued Mumford with only slight exaggeration, "is nothing less in fact than the concrete model of the conceptual world which was built up by the physicists of the seventeenth century."[46]

No less important, Mumford goes on to make two further points: (1) that "the practices of the mine do not remain below the ground but...affect the miner himself and...alter the surface of the earth"[47]; and (2) that "more closely than any other industry, mining was bound up with the first development of modern capitalism"[48]-- including its role in forging the sinews of war for its exploitative adventures around the globe. In the context of the acid rain debate, these two points from ecological and political eonomic perspectives reveal the extent to which appropriate environmental policy is intimately linked up with sound energy policy and a democratic

democratic economy. Thus, exploitative appropriation of the earth's subcutaneous resources damages the tissues of its outer skin, its human hosts, and the surrounding life-sustaining environment.

CONCLUSION

Acid rain then is not an isolated, discrete issue for public policy. Rather it is intertwined in a network of other energy, environmental, and health issues which must be addressed in their complex and dynamic interaction. Consequently, the problem of acid desposition will not go away if it is conceptualized and responded to in a single-focused, narrow, technical manner. The central task of critical policy analysis and innovative political action is to break through the supposed policy paradoxes and energy-environment-economic policy "trade-offs" which have been structured into our political thinking by the interest-group framework of the middle-levels of power in America and into our philosophical outlook by the antagonistic dualisms of modern Western thought and culture.

In perhaps more concrete terms, I am suggesting the need for a broader multi-class, multi-racial movement for ecological repair, a strategy which moves from pursuit of legalistic remedies to the quest for political actions involving the activation and empowerment of local citizens in environmental disputes, a program which bridges class interests and relates ecological repair to a democratic economy, and an orientation which goes beyond the "not in my backyard"/obstructionist view of the past to a positive, reconstructive posture suitable to a future society where corporate greed and interest-group selfishness have been overcome. The political maturity of the American environmental movement, perhaps inspired by the example of the West German Green Party and the evolving Green movement wordwide, would be a development of no small moment not only for the improvement of the environmental health and public safety of the American working class and wider public, but also for the potential evolution of a new social order and a political economy based upon democratic values and ecological imperatives.

NOTES

1. See the essays in the collection edited by John Baden--Baden, ed., **Earth Day Revisited** (Washington, D.C.: The Heritage Foundation, 1980). See, too, Robert J. Smith, "Privatizing the Environment," **Policy Review**, No. 20 (Spring 1982), pp. 11-50; Smith, "Conservation and

Capitalism," **Libertarian Review,** October 1979, pp. 19-25; Robert Tucker, "Is Nature Too Good for Us?" **Harper's,** March 1982, pp. 27-35; and Tucker, "The Environmental Era," **Public Opinion,** February/March 1982, pp. 41-47.
2. Consider the critical analysis of America's political economy in: Herbert G. Reid and Ernest J. Yanarella, "Beyond the Energy Complex: The Role of the Public Sphere," a paper prepared for the annual meeting of the American Political Science Association, New York City, August 30-September 3, 1981.
3. Leo Marx, "American Institutions and Ecological Ideals," **Science,** Vol. 170 (November 27, 1970), p. 945. For a recent interview of Marx which deals with further thoughts on pastoralism in literature and politics and, in particular, his developing conception of a "middle landscape," see: Bruce Piasecki, "Pastoral Ideals and Environmental Problems: An Interview With Leo Marx," **The Amicus Journal,** Fall 1982, pp. 56-61.
4. On the philosophy of ecology, a number of recent works are illuminating: Bill Devall, "The Deep Ecology Movement," **Natural Resources Journal,** Vol. 20 (April 1980), pp. 299-322; Murray Bookchin, **Toward an Ecological Society** (Toronto, Canada: Black Rose Books, Ltd., 1980); Donald Wooster, **Nature's Economy: The Roots of Ecology** (Garden City, N.Y.: Anchor Books, 1979); and Jonathon Porritt, **Seeing Green: The Politics of Ecology Explained** (New York: Basil Blackwell, 1985).
5. Craig Humphrey and Frederick Buttel, "The Environmental Movement: Historical Roots and Current Trends," **Environment, Energy & Society** (Belmont, Mass.: Wadsworth Publishing Company, Inc., 1982), pp. 112-119; and Allan Schnaiberg, **The Environment: From Surplus to Scarcity** (New York: Oxford University Press, 1980), pp. 367-368, 381-382, especially.
6. Samuel P. Hays, **Conservation and the Gospel of Efficiency: The Progressive Conservation Movement, 1890-1920** (New York: Atheneum Books, 1969), p. 263.
7. Schnaiberg, **The Environment,** p. 382. For an anlaysis of the Sierra Club and other preservationist groups, see Jerome Price, "Environmentalist Antinuclear Groups," **The Antinuclear Movement** (Boston, Mass.: Twayne Publishers, 1982), pp. 39-64, esp. pp. 43-49; and Stephen Fox, **John Muir and His Legacy** (Boston: Little, Brown and Company, 1982).
8. Lorna Salzman, "Ecology and Politics in the United States, **The Ecologist,** Vol. 12 (1983), p. 41.
9. Schnaiberg, **The Environment,** p. 382. He goes on to acknowledge, however, that "the two now subsumed a broader range of ecosystem dimensions, particularly air pollution." (p. 382)
10. Douglas and Wildavsky highlight this point in **Risk and Culture,** pp. 129ff.
11. Sheldon Wolin makes this same critical observation about the impact of various peace organizations, like

Ground Zero, on public discussion on the nuclear war issue in: Wolin, "Editorial," **Democracy**, Vol. II (July 1982), pp. 2-4.
12. See, again, the valuable, but overcategorized, distinctions between hierarchical and decentralist evironmentalist organizations in Douglas and Wildavsky, **Risk and Culture**, pp. 127-151; and Price, **The Antinuclear Movement**, pp. 49-52. Both deal with the Friends of the Earth.
13. Douglas and Wildavsky, **Risk and Culture**, pp. 128-129.
14. Walter Dean Burnham speculates on the long-term repercussions of interest-group syndicalism on America in his book, **The Current Crisis in American Politics** (New York: Oxford University Press, 1982).
15. This apposite phrase is Claus Offe's--Offe, "'Reaching for the Brake': The Greens in Germany," **New German Critique**, No. 11 (Spring 1983), p. 51.
16. Schnaiberg, **The Environment**, p. 374.
17. **Ibid.** In a critique which converges with Schnaiberg's, two other environmental sociologists argue that "the failure of these two groups [environmentalists and "underdogs"] to reach these understandings in turn may lead to the rightist state and its inegalitarian, authoritarian style of 'environmental adaptation'"--Frederick Buttel and Oscar Larson III, "Whither Environmentalism? The Future Political Growth of the Environmental Movement," **Natural Resources Journal**, Vol. 20 (April 1980), pp. 343-344.
18. Salzman, "Ecology and Politics in the United States," p. 40.
19. See James M. Markham, "In a 'Dying' Forest, the German Soul Withers Too," **New York Times**, May 25, 1984, p. 2.
20. Morris Berman and Capra and Spretnak critically comment on the eco-fascist legacy and the need and efforts of ecologists to overcome it--Berman, **The Reenchantment of the World** (New York: Cornell University Press, 1981), pp. 290-292; and Capra and Spretnak, **Green Politics: The Global Promise** (New York: E.P. Dutton, Inc., 1984), p. 14.
21. For a lucid and insightful treatment of the SPD and the rise of parliamentary and extra-parliamentary opposition in West Germany to its "success," see Joyce Marie Mushaben, "The Changing Structure and Function of Party: The Case of the West German Left," **Polity**, forthcoming.
22. **Ibid.**, p. 23.
23. Wolfgang Rudig, "The Greening of America," **The Ecologist**, Vol. 13 (1983), p. 37. See, also, Horst Mewes, "The West German Green Party," **New German Critique**, No. 28 (Winter 1983), pp. 51-85, esp. 53-63.
24. In addition to Capra and Spretnak's work, see Elim Papadakis' more journalistic study, **The Green Movement in West Germany** (New York: St. Martin's Press, 1984).

25. Capra and Spretnak make this key point, but fail to connect it to the broader "post-modern" turn in phenomenological, western Marxist, and critical theoretical traditions which emerged in twentieth-century philosophy in continental Europe--Capra and Spretnak, **Green Politics**, pp. xviii-xx.
26. For a synoptic view of this and other themes bearing on the crisis of the modern philosophical worldview and on philosophical and cultural perspectives attempting to sublate its shortcomings, see Herbert G. Reid and Ernest J. Yanarella, "Toward a Post-Modern Theory of American Political Science and Culture: Perspectives from Critical Marxism and Phenomenology," **Cultural Hermeneutics**, Vol. II (August 1974), pp. 91-166; and Reid and Yanarella, "Political Science and the Post-Modern Critique of Scientism and Domination," **The Review of Politics**, Vol. 37 (July 1975), pp. 286-316.
27. Besides the above two essays, two books are recommended: Fred Dallymayr, **Twilight of Subjectivity: Contributions to a Post-Individualist Theory of Politics** (Amherst, Mass.: University of Massachusetts Press, 1981); and Morris Berman, **The Reenchantment of the World**.
28. Die Grunen, **Programme of the German Green Party** (London: Heretic Books, 1983), pp. 7-9.
29. Offe, "'Reaching for the Brake'," p. 46.
30. Issues discussed in Capra and Spretnak, **Green Politics, passim**.
31. On the West German policy turn-about on multi-lateral agreements on curbing transfrontier acid rain damage, see: Bette Hileman, "1982 Stockholm Conference on Acidification of the Environment," **Environmental Science & Technology**, Vol. 17 (January 1983), pp. 15A-18A, and Sweden, Ministry of Agriculture, **The 1982 Stockholm Conference on Acidification of the Environment, June 21-30, 1982-- Proceedings** (Stockholm, Sweden: Departementens Reprocentral, 1983). The policy actions of the CDU on acid rain control are reported in: Elizabeth Pond, "West Germans Discover Their Forests are Dying, In Response, Bonn's Conservatives Grab Ecology Issue from Greens," **Christian Science Monitor**, August 24, 1983, p. 3; Pond, "Exhausted But Clean: Germany Goes Unleaded," **Christian Science Monitor**, July 21, 1983, p. 18; Pond, "Speed Limits in Germany--Are You Joking?" **Christian Science Monitor**, November 1, 1983, p. 15. On acid rain and the dying of the German forests, the following articles may be consulted: Siegfried Niebuhr, "West Germany Trying to Save Forests from Damage by Gases," **Los Angeles Times**, June 18, 1983, Pt. 1-A, p. 2; "West Germans Fear a Calamity as Acid Rain Damages Forests," **Washington Post**, December 28, 1983, p. 14; and "Wir stehen vor einem okologischen Hiroschima," **Der Spiegel**, February 14, 1983, pp. 72-73, 76, 79, 81, 84-85, 87-89, and 92.
32. For treatment on the specific features of the Green program on acid rain, see Capra and Spretnak, **Green**

Politics, pp. 33 and 34; Pond, "Exhausted But Clean," p. 18; Pond, "Speed Limits in Germany," p. 15; "Wir stehen...okologischen Hiroschima," pp. 72-73; and Horst Mewes, "Can Germany's Greens Succeed?" **Environment**, Vol. 25 (May 1983), p. 4.
33. Capra and Spretnak, **Green Politics**, pp. 87-97; Die Grunen, **Programme of the German Green Party**, pp. 9-14; and Mewes, "Can Germany's Greens Succeed?" p. 2.
34. Capra and Spretnak, **Green Politics**, p. 28. Consider, in this light, Alan Wolfe's essay, "Why is There No Green Party in the United States?" **World Policy Journal**, Vol. I (Fall 1983), pp. 160-180.
35. Robert Boyle and R. Alexander Boyle, **Acid Rain** (New York: Schocken Books, 1983), p. 12.
36. Office of Science and Technology Policy, "Interim Report...," p. 2. Later on, they reinforce this point with the comment: "if we take the conservative point of view that we must wait until the scientific knowledge is definitive, the accumulated and damaged environment may reach the point of 'irreversibility'." (p. 5)
37. J.E. Lovelock, **Gaia: A New Look at Life on Earth** (New York: Oxford University Press, 1979).
38. **Ibid**, p. ix. Although Lovelock's political conservatism blunts the radical implications of some of his more insightful points, his book has virtually effected a paradigm shift in some scientific subcommunities--see, e.g., Dorion Sagan and Lynn Margulis, "The Gaian Perspective of Ecology," **The Ecologist**, Vol. 13 (1983), pp. 160-167; and J. Donald Hughes, "GAIA: An Ancient View of our Planet," **The Ecologist**, Vol. 13 (1983), pp. 54-60.
39. For a more productive, yet ambiguous, concept of Nature drawn from the medieval worldview, see Carolyn Merchant's **The Death of Nature: Woman, Ecology and the Scientific Revolution** (New York: Harper & Row, Publishers, Inc., 1980).
40. Lovelock worries about the strategic vulnerabilities found in the more elemental building blocks of Gaia in the soil and the deep sea--Lovelock, **Gaia**, pp. 113-114, 117, and 121. I have noted certain parallels between Lovelock's concerns and science fiction writers' fears of eco-catastrophe in: Ernest J. Yanarella, "Slouching Toward the Apocalypse: Visions of Nuclear Holocaust and Eco-Catastrophe in Contemporary Science Fiction," a paper prepared for the annual meeting of the American Political Science Association convention, Washington, D.C., August 30-September 2, 1984.
41. Office of Science and Technology Policy, "Interim Report from OSTP's Acid Rain Peer Review Panel," (Washington, D.C.: Press Advisory, June 28, 1983), p. 4.
42. See, generally, Wetstone and Rosenkranz, **Acid Rain in Europe and North America**, op. cit.; Ellis Cowling, "Acid Precipitation in Historical Perspective," pp. 113A-116A; Environmental Resources Limited, **Acid Rain: A**

Review of the Phenomenon in the EEC and Europe (New York: Unipub,1983); and "East and West Agree in Munich on Pollution," **German Tribune**, July 15, 1984, p. 12.

43. Ellis Cowling documents this point in his historical review of the transformation of acid rain from a scientific problem to a public issue--Cowling, "Acid Rain in Historical Perspective," **passim**.

44. W. Herbert, "Long Hot Future: Warmer Earth Appears Inevitable [NRC and EPA reports]," **Science News**, Vol. 124 (October 22, 1983), p. 260; and R.A. Kerr, "Carbon Dioxide and Changing Climate," **Science**, Vol. 222 (November 4, 1983), p. 491.

45. Merchant, **The Death of Nature**, chs. 1 and 2; and Lewis Mumford, **Technics and Civilization** (New York: Harcourt, Brace and Company, 1934), pp. 60-81.

46. Mumford, **Technics and Civilization**, p. 70.

47. **Ibid.**

48. **Ibid.**, p. 74.

10
Acid Rain—Acid Diplomacy

John E. Carroll

INTRODUCTION

The geological vulnerability of a nation's land and water areas has much to do with the diplomatic behavior of that nation relative to acid rain and international relations. The economic dependency of a nation on the pollution emissions which cause acid rain, especially with respect to sulfur dioxide (SO_2) emissions from coal and oil-fired electric power plants and metal smelters, likewise directly influences a nation's diplomatic behavior. Some nations and, in the case of the U.S., some states, are emitters of acidic pollution and, if they are upwind, generally behave as emitters. Others, generally downwind, are receptors of damage from such transborder emissions, and behave as such. There are some nations which remain neutral or disinterested, at least at the present time, but this category is declining. There is, therefore, a political geography of acid rain which underlies all international diplomacy and attempts at such in this field.

Acid rain in Europe and North America represents the kind of international imbalance or asymmetry which guarantees the existence of a serious transborder diplomatic problem. In Europe a geologically well-buffered and coal-burning-dependent United Kingdom lies immediately upwind of a very vulnerable and a relatively non-polluting Norway and Sweden. An excessively high sulfur coal-dependent eastern Europe lies too often upwind of alpine forests of Germany, Switzerland and Austria and, as well, the vulnerable plateau of Scandinavia. To reduce pollution emissions from the U.K. or eastern Europe is a costly enterprise of little perceived benefit to countries struggling to rise above economic chaos, and damaged recipients of this pollution can do little other than to raise the cry in international fora and in the broader community of nations--an idealistic call too often falling on deaf ears.

In North America, a high-sulfur coal-dependent industrial heartland in the Midwest, a region saddled with its share of economic recession and itself ecologically well-buffered, lies upwind of a very vulnerable and less coal-dependent Canada, causing the latter to join Scandinavia and other damaged nations in the call for surcease through controls on emissions. This has led to the development for Canada of a serious bilateral dispute with the United States which, while not without its own domestic concerns over acid rain (especially in the Northeast), nevertheless identifies, as does the U.K. and eastern Europe, with emission-dependent interests. With such asymmetry on two continents, acid rain is today in the forefront as an issue in international environmental diplomacy.

THE EUROPEAN SCENARIO

The peoples of northern Europe have been concerned about environmental effects of acidic precipitation for at least a decade longer than those of North America. Much of Scandinavia in particular is geologically vulnerable to acid deposition, with little buffering capacity in its granitic bedrock base, much like the Canadian Shield. The region is also downwind of the heavily industrialized United Kingdom and Germany. It is thus not surprising that Norway, Sweden and Finland have felt and continue to fear the effects of acid rain on their lakes, aquatic ecosystems and, potentially, on their important commercial forest resources. There is a direct parallel between vulnerable Scandinavia and the industrialized United Kingdom and Germany, on the one hand, and vulnerable Canada and the more industrialized United States, on the other.

Scientific research began early in Scandinavia; the problem became well known; and support developed for political initiatives. The British and, prior to 1982, the West German reaction to those Nordic initiatives has not been unlike the U.S. reaction to Canada's complaint: a call for more research before expensive action is taken. Indeed, there is a further parallel in that the level of awareness and public knowledge is much greater in the receptor nations on both sides of the Atlantic than it is in the emitter nations.

With many more sovereign nations and international borders over a small land area, the necessity for and interest in international arrangements or accommodations of some form is naturally greater in Europe than in North America. Thus the history of attempts to reach multilateral agreements governing long-range transport of air pollutants, including acid

rain, is more substantial in the former than in the latter.
The European experience with the international relations aspects of acid rain centers on the efforts of the Economic Commission for Europe (ECE), the Council of Europe, the European Community (EC), the Organization for Economic Cooperation and Development (OECD), and various arrangements between the Nordic nations (Norway, Sweden, Finland, and Denmark). The United Nations Environment Program (UNEP) has devoted limited attention to atmospheric pollution, but this has been concentrated on third world nations rather than on Europe or North America and has been globally oriented.[1]

The **Economic Commission for Europe (ECE)**, one of five regional economic commissions of the United Nations, established a Working Party on Air Pollution Problems in 1969 and a Special Group on Long-Range Transboundary Air Pollution in 1978. This group meets regularly and has established a uniform monitoring and data evaluation program throughout much of Europe. Perhaps most notable, the ECE hosted a high-level meeting in November 1979, which drafted a Convention and Resolution on Long-Range Transboundary Air Pollution. This convention, signed by Canadian Environment Minister John Fraser and U.S. EPA Administrator Douglas Costle as well as European environment ministers, required the contracting parties:

--to limit and gradually reduce and prevent air pollution, including LRTAP;

--to develop policies and strategies to combat the discharge of air pollutants;

--to exchange information and review their policies and consult on request;

--to initiate and cooperate in the conduct of research and the development of technologies for reducing emissions of sulfur compounds and other major air pollutants, techniques and models for assessment as well as education and training programs.

The convention does not bind signatory nations and does not provide supervisory mechanisms. Nor does the convention rule on state liability as to damage. It aims more to manage LRTAP than to provide a remedy for injured states or individuals. It was ratified by the requisite 24 nations and entered into force in March of 1983.

As might be expected, Norway and Sweden--two of Europe's most geologically and ecologically vulnerable nations--were among the leaders in drafting the ECE convention. Norway, in particular, has hopes that this convention will not be limited to acid rain but will be an instrument to deal with many other international upper atmospheric problems, including the question of ozone as a transborder pollutant.[2]

The **Council of Europe**, founded in 1949 to achieve greater European unity, adopted a recommendation on air pollution by sulfur emissions in March 1970 calling for measures to reduce sulfur particles in combustion gases and governmental coordination of efforts on land planning and air pollution. A 1971 resolution of the council recommended that governments grant residents of border regions of adjacent nations the same protection they grant their own inhabitants. Van Lier has concluded that the council's chief role has been to serve as a laboratory of ideas, initiating and stirring up interest and then letting organizations with greater competence assume the real work.[3]

The **European Community (EC)** has a general involvement in environmental matters within a broader mandate to achieve a constant improvement of the living and working conditions of their peoples. The EC is particularly interested in pollution problems arising in certain industrial sectors and energy production and is especially concerned with industries emitting dust, SO_2 and NO_x, hydrocarbons and solvents, fluorine, and heavy metals. Its work revolves around organized exchanges of information, establishment of administrative and scientific bodies responsible for air management, use of economic measures, establishment of standard monitoring networks, and possible effects of transfrontier pollution. Van Lier notes that the EC provides a forum for affected nations such as the Netherlands to communicate with net polluters such as France and West Germany.[4] However, the absence of Norway and Sweden from EC membership reduces its value in solving acid rain problems.

The **Organization for Economic Cooperation and Development (OECD)** was founded in 1961 to promote ecnomic growth and raise living standards for its member nations. Its membership includes most European nations (except East European states), the United States, Canada, and a few other nations. In June 1974, OECD adopted "Guidelines for Action to Reduce Emissions of Sulphur Oxides and Particulate Matter from Fuel Combustion in Stationary Sources," which set a number of objectives on clean fuels and sulfur content in fuels. In June 1974, OECD called on member governments to reduce emissions of SO_2 and particulates, to develop measures for reducing emissions of NO_x and hydrocarbons, to encourage emission monitoring, and to assess the effects of acid deposition. The organization has been especially concerned with the environmental impacts of energy generation, and it advocates the "polluter pays" principle as an economic incentive technique (i.e., payment for a right to pollute). OECD has also urged cooperation in the development of international law applicable to transfrontier (or transborder) pollution and since 1974 has issued many recommendations to achieve various aspects of this objective. In 1979, it adopted specific recommendations on coal burning aimed at reducing environmental impacts. In addition to its intense interest in the reform of international environmental law and the linkage between economics and environmental control, OECD has launched a number of research projects dealing directly with acid rain. Chief among these was the 1970-73 study on Air Pollution from Fuel Combustion in Stationary Sources[5] and the 1972-79 cooperative technical program to measure the long-range transport of air pollutants, which has produced one of the leading documents regarding scientific information on LRTAP.

Norway, Sweden, Finland, and Denmark are among the countries most threatened by acid rain and also among the most unified groupings of European nations. In 1974, they signed the **Nordic Convention on the Protection of the Environment,** which lays a strong foundation of regional unity to respond to this common problem. The convention largely eliminated the effect of international boundaries within the Nordic nations and provides reciprocal environmental protection to all citizens of the four nations from threats from within. The similarity of domestic legal structures and approaches within these countries has enabled a high degree of integration and coordination and has ensured a strong regional response to external environmental threats. The extent of the ambitious scientific research programs of Norway and Sweden, small nations by any standards, on the nature and effects of acid rain, demonstrates the seriousness with which these two countries regard the problem. Further evidence of their deep interest and concern over LRTAP is the extensive development of strategies and policies designed by their governments to achieve abatement of internal air pollutant emissions, particularly sulfur, and the strong interest of their environment and foreign relations ministries to establish international accords and arrangements to reduce external sources. Therefore, it is logical to look to such sources of interest for guidance in developing international arrangements and achieving diplomatic solutions to the acid rain issue in North America.

THE CANADA-UNITED STATES SCENARIO

Although involving only two nations, the North American situation appears no less complex, no less inscrutable, than the European condition. U.S. dependence on very considerable high sulfur coal reserves, their mining and conversion into electricity, and the use of that electricity to fuel the declining but economically and politically important industrial heartland of the Midwest, and the lack of viable alternatives for equivalent energy production, ensures the U.S. position of recalcitrance before demands, domestic and international, to reduce SO_2 emissions. This is in spite of damage in the Northeast and political calls for action from that region. Extremely high potential for damage north of the border in Canada, the downwind location of that country relative to U.S. emissions and as well Canada's lack of dependence on coal for electricity, ensures that country's position and interests as contrary to those of the U.S. Canada is, however, a major generator of SO_2 due to its very large sulfur-emitting metal smelters in the North, as well as its high sulfur Nova Scotia coal, and is thus forced realistically to "bite the bullet" and work to keep its own house in order. Its real incentive to do this, however, is diminished somewhat by U.S. reluctance to take action.

The problem has become a serious one for two basic reasons. Canada has unilaterally raised the acid rain issue and given it a high place on the diplomatic agenda, and the United States has thus far failed to respond substantively to this concern, sending a very negative signal back to Canada and its people.

THE CANADIAN CONCERN

Why has Canada raised the issue to such a high diplomatic level and placed such a high political stake on its resolution?

First, at least 50% of all of Canada's acid deposition (and perhaps as much as 60%) comes from U.S. sources over which Canada exerts no influence, except through diplomacy. Conversely, only a very small percentage of U.S. acid deposition (perhaps 15%) comes from Canadian sources. This fact has been determined by a group of highly qualified government scientists from both Canadian and U.S. federal governments, the Bilateral Research Consultation Group. Thus, there is international movement of pollutants across the border in both direction, but the national contributions to

that transboundary movement are not close to being
balanced. All interests accept this as fact. Hence,
no matter what sacrifices Canada makes to reduce its
own emissions, it still faces very substantial
pollution over which it has no control.

Second, a much higher proportion of Canada is
geologically vulnerable to the impacts of acid
deposition than is the U.S. Specifically, almost all
of Atlantic Canada, Quebec and Ontario, most of
Manitoba and Saskatchewan, and smaller portions of
Alberta, British Columbia and the northern territories
of Canada are highly vulnerable. This vast region,
much of which is known as the Canadian Shield, is
composed of exposed granite bedrock, already highly
acidic soils, numerous small lakes, and vast coniferous
forests, none of which can take any additional incoming
acid without experiencing serious ecological change.
Not the least of these changes is the death (for all
intents and purposes) of thousands of small lakes,
bodies of water which can no longer support fish or
other aquatic life. In other words, only small areas
of Canada have buffering capacity sufficient to accept
this acidic deposition without showing signs of damage,
at least in the short-term. Areas such as the prairie
grasslands, the farmlands of southwestern Ontario and
Prince Edward Island, and a few other Canadian areas
are in this buffered category. Further exacerbating
the situation, however, is the fact that some of
Canada's most cherished lake and recreational resort
country is in the zone of highest vulnerability. This
has been described in the literature and painting of
Canada's most revered writers and artists and is
located just north of Toronto and other major cities
and heavily used by thousands of Canadians--the Muskoka
Lakes and Haliburton Highlands. Thus, the proximity of
acid rain damage (or alleged acid rain damage) to large
numbers of people, including the most influential
decision-makers of Canadian society, has contributed to
the seriousness of the bilateral problem.

Third, Canada is, perhaps more than any other
nation, dependent upon the forest products industry for
export of wood products to much of the world. Many
jobs are dependent on Canada's lumber, pulp and paper,
and newsprint exports and on the country's ability to
sustain a competitive position on world markets. What
does this have to do with acid rain? The data are by
no means complete, but there is increasing suspicion
among government and industry foresters and among
academic scientists that acid deposition may be
inhibiting the growth of commercially valuable softwood
timber. The full story is yet to unfold, but Canadian
forestry officials believe that a decline in growth
rate of only 1% will be sufficient to drive Canada off

the world markets--especially when combined with
Canada's other disadvantages of a short growing season,
high labor costs, and increasingly sharp competition
from the American Southeast. Anything which threatens
or appears to threaten Canada's world position in
forestry exports must be viewed critically by any
Canadian government. Many are now awaiting research
results in this area with great anticipation, not the
least being the forest industry and those regions
dependent on it.

A final factor contributing to the seriousness of
acid rain as a bilateral issue is the great disparity
between the peoples vis-a-vis their knowledge of acid
rain, and concern over its effects. Most Canadians
have been saturated with media attention on this
subject for at least four or five years, and polls
indicate that as many as 85% of all Canadians
nationwide are well aware of the issue. Torontonians,
residents of Canada's largest city, are especially
aware and have heavily influenced their politicians in
Ottawa to negotiate a hard line with the U.S. Americans
have had significantly less exposure to the issue in
general and to its impact on bilateral relations in
particular. Polls indicate that American awareness was
until recently the obverse of Canadian awareness--in
other words, only 15-20 percent. It is climbing,
however, and may be as high as 50 percent today. This
differential goes a long way toward explaining such an
obvious lack of concern south of the border. The
disparity in awareness is dangerous in the bilateral
sense, for the image conveyed to Canadians is that
Americans simply don't care, that they willingly reap
the material benefits of uncontrolled pollution
emissions while assuming no responsibility for serious
damages to Canada. Further, as signatories to a joint
Memorandum of Intent (in 1980) to resolve the problems,
Americans do not seem to be taking such obligations
seriously. The diplomatic damage which results from
this perception, whether justified or not, is a threat
to future U.S. interests in Canada as well as to
Canada's long-term well being.

These four imbalances all establish the foundation
for fundamental asymmetry which should alone be
sufficient to cause a serious bilateral problem.

As a transboundary environmental problem, as a
problem of international environmental diplomacy, in
North America and in Europe, acid rain is of a much
greater order of magnitude than the more long-standing
problems of air and water pollution at the border,
disputes over internationally shared river-basins, etc.
Many, many millions of dollars, thousands of jobs and
whole regional--if not national--economies can be
affected by decisions in this area in contrast to the

much smaller stakes of typical upstream-downstream, or upwind-downwind problems at the border. Perhaps for the first time afflicted countries are calling upon neighboring nations to make changes in their habits deep within their national territory (rather than near to the border), and to make changes of a much greater order of magnitude than ever before. The relatively primitive regimes of international law, custom and policy and the institutions designed to implement such law, custom and policy are strained to capacity and beyond in trying to cope with this problem, whether in Europe or North America. And the problem is not likely to be contained on those two continents for it already threatens rapidly industrializing regions of the lesser developed nations in both northern and southern hemispheres. The frequency and intensity with which this issue is raised worldwide by diplomats of damaged nations, by scientists of all nations (and of many different disciplines), coupled with the increasing concern now being shown in the SO_2 emitting regions in the U.S., U.K., eastern Europe and elsewhere, portend a continued focus on this subject bilaterally and in multilateral international fora--such as the Organization for Economic Cooperation and Development, the European Community, and in the UN and its agencies--for some time to come.

TOWARD RESOLUTION

Obviously a bilateral diplomatic resolution of this problem between any two nations could be achieved by the linkage of unrelated issues and the establishment of a quid pro quo along political, economic or other lines in as many innovative ways as the mind can devise. To hypothesize on such, however, would be of no value here. What might be done to resolve the Canada-U.S. acid rain impasse, an impasse dating from the breakdown of negotiations in July 1982? Fortuitously, such an effort would also alleviate--if not resolve--the problem domestically within each country.

Canada has called upon the United States to join it in a bilateral reduction of SO_2 emissions in each country in a ten-year period, using 1980 as the base year against which to measure reduction in actual emissions. However, Canada is a leader in Europe and in multilateral fora (and is joined strongly by the Scandinavian nations, since 1982 by the Federal Republic of Germany, and by others in Europe) in its call for a 30% reduction in emissions over ten years. Given the U.S. recalcitrance to build a large number of costly wet sulfur scrubbers (the only real way to achieve a 50% reduction at this time), Canada's bilateral plea for a 50% reduction will continue to fall on deaf ears. But, can the U.S. achieve the very respectable 30% reduction in that period, a reduction which would be lauded in Europe and one which Canada would have to accept, given its well known charter membership in the "30% Club of Nations"? Available evidence indicates that the U.S. could well achieve such a reduction at significantly less expenditure (i.e., without sulfur scrubber technology) and, in fact, in a time period of much less than ten years, a fact presumably of some interest to American environmentalists and to residents of the afflicted Northeastern states.

PRESCRIPTION: THE 30% SOLUTION

Both sides in the now well-known acid rain debate have become so polarized that resolution is becoming a most illusory, if indeed, a still possible goal. Coal producers, coal-dependent utilities, coal-dependent regions like the Midwest, metal smelters and other significant emitters of sulfur dioxide gases are shouting loudly and vigorously that the scientific evidence does not prove their emissions are guilty of the considerable and irreversible damages claimed by their adversaries. Environmentalists, New Englanders, Canadians, and others residing in or otherwise concerned about highly vulnerable pollution-receiving environments downwind of the emitters are just as loudly shouting for protection from those emissions, for a surcease in the acidity of precipitation falling upon their lakes and forests.

One side says no cause and effect relationship is provable, that scientific evidence is lacking, that little is known, that more research is needed, that emission reductions are so costly as to threaten regional economies and employment in the coal-dependent regions, and that no action is justified at this time. The other side says that the cost of enduring the ecological damage more than justifies the needed investment in emission controls, that this damage is irreversible, and that we have the technology to make significant emission reductions now and must proceed forthwith toward at least a 50% sulfur dioxide reduction in the next ten years regardless of cost. The result: political stalemate, significant and damaging regional rivalry, and, since a vulnerable Canada is downwind of and receives a great deal of U.S. pollution emissions blowing across the border, long-term damage to Canadian-American relations--damage of a much more serious nature than most Americans realize.

Can we resolve the matter and do so now? Yes, if we have the will. Good-faith compromise is needed and needed today.

Pollution emitters must come to realize that the weight of scientific evidence, as put forth by qualified credible scientists of so many different scientific disciplines in Europe, Canada and the United States, can no longer be ignored. While "proof," as most Americans use that word, may not exist (and may never exist), there is now more than enough strongly held suspicion that can no longer be discounted and that must command respect in any rational person. To argue against these widely-held scientific views is no longer a defensible position and should be abandoned by those who would hope to preserve their own credibility. The evidence is there--the time for response through tangible emissions reductions has arrived.

But how much reduction? A case can be made that a 30% SO_2 emissions reduction over the next ten years is a most reasonable compromise. It is the level of reduction espoused by the so-called "30% Club," those nations opting for a 30% reduction in SO_2 emissions over the next ten years. Membership includes the Federal Republic of Germany, France, Sweden, Norway, Canada and others, and the list is growing. Even the East European nations and the Soviet Union are now included, at least in terms of 30% reduction of transborder emission, if not 30% of all emissions.

The 30% reduction level is not only most respectable as a tangible good-faith compromise, but it is attractive for another important reason: it is achievable without installing extremely expensive capital-intensive technology such as wet sulfur scrubbers.

Herein lies the other side of the coin. Environmentalists and others calling for sharp emissions reductions must begin to acknowledge the proven costliness of their demands, as well as the irreversible capital commitments which would have to be made if sulfur scrubbers, necessary to achieve 50% reductions, are to be more widely built. Advocates of these reductions must also come to realize that there is a profound impact on high-sulfur, coal-dependent regions such as the Midwest, that sharp reductions can cause much increased unemployment in the already hard-hit coal mining regions, and that loss of competitiveness in the energy intensive coal-dependent industrial heartland will result. Just as environmentalists argue that damage to lake and forests must not be taken lightly, they must also recognize that this kind of threat to the coal-dependent regions must also not be taken lightly.

The answer: approval today of 30% reduction over

the next ten years. The means for achieving such a reduction, though, would not be through the construction of costly sulfur scrubbers which are so capital intensive as to preclude many other options. Rather, the means would involve: the highly conventional, significantly less expensive, and infinitely more flexible techniques of universal and mandatory coal pre-washing at the mines; the shifting of baseload utility-generating capacity onto newer less polluting plants, reserving the older polluters for peak-load generating only (this is a utility operational technique known as least-emission dispatching and contrasts with the present technique of least-cost dispatching); and the adding of increased lower sulfur southern and western coal into the fuel mix, not so large an admixture as to threaten the viability of the high sulfur coal industry (an industry which must be protected in the national interest), but enough simultaneously to reduce emissions and to give incentive to our low-sulfur coal producers to become more involved in the resolution of the problem.

Can we achieve a 30% reduction in this way without costly scrubbers or other big expenditures of capital? Yes. And if environmentalists are serious about achieving a resolution, they will recognize that they need not wait ten years to achieve this reduction; for ten years is not necessary to carry out these techniques--they can be done much more rapidly. In the meantime, research into acid rain can and must continue, and all decisions can be reassessed at any point in the future depending on the results of this research. No irreversible decision will have been made.

If utilities, the coal industry and environmentalists truly want resolution, they will start the process now. The ball is in their court. Threatened Canadians, New Englanders, Midwesterners, and all other interested North Americans (and Europeans) will be watching anxiously. Who will begin the process? Who will be the first to deserve respect for launching a good-faith effort?

Why has this solution not been adopted? A major reason lies in the U.S. failure to value sufficiently its foreign relations and particularly its relationship with Canada, which is an important bilateral relationship--albeit one which has long been taken for granted south of the border. The distaste of the present U.S. administration for government regulation in general, environmental or otherwise, the determination of that Administration not to spend tax dollars in such control efforts, and the political power of the U.S. Midwest (of both major parties as this has become a nonpartisan issue) to maintain the status quo are all important factors in the lack of U.S. action to date. As scientists and national governments continue to move toward supporting action today rather than later, however, the "do nothing but conduct more research" alternative will become increasingly indefensible and untenable. As nations one after another opt for some action now while more research is conducted and evidence gathered, nations which take no action will find themselves more and more in the spotlight. The United States and the United Kingdom are very much in that spotlight today.

The cost in damage to the Canada-U.S. relationship is admittedly difficult to measure with any precision, but it is there. It is real. It is unnecessary. Canada and the United States, two North American neighbors, have the opportunity to show Europe and the rest of the world how a difficult and seemingly intractable international environmental problem can be resolved and removed from the agenda of differences afflicting nations. Only the will to act stands in the way of bilateral cooperation.

NOTES

1. A survey of this subject can be found in **Acid Rain and International Law**, by Irene H. Van Lier, L.L.M., Bunsel International Consultants (Toronto, Canada, and Sijthoff and Nordhoff International Publishers, Alphen Aan Den Ryn, The Netherlands, 1980). A further treatment of the international legal aspects of acid rain may be found in **World Public Order of the Environment**, by Jan Schneider (University of Toronto Press, 1979).
2. Rolf Hansen, Norwegian Minister of the Environment, Statement in the Generla Debate, ECE High Level Meeting on the Environment (Geneva, Novermber 12-15, 1979).
3. Van Lier, **Acid Rain and International Law**, p. 153.
4. Ibid., p. 160.
5. OECD Environment Directorate, **Report and Conclusions of the Joint Ad Hoc Group on Air Pollution from Fuel Combustion in Stationary Sources** (Paris: OECD, 1973).

Conclusion:
A Prospectus on the Future

The future course of the acid rain debate remains clouded by the redundant nature of the policy stalemate within the legislative, administrative, and scientific arenas. Earlier parts of this volume have explored the uniqueness of the acid rain debate in comparison with previous environmental disputes, the varying policy alternatives currently being generated by researchers, policy analysts, and politicians concerned with overcoming or finessing the policy immobilism on the acid rain front, and the practices and strategies of other political actors for addressing the effects of acid deposition.

In this concluding part, the chapter by George Freeman presents a prospectus on the future of the debate over acid rain control, first, by reviewing the complex and multi-dimensional character of the policy struggle and, second, by sifting through the scientific uncertainties and political conflicts in search of the possible ingredients for reaching political consensus on a workable and acceptable approach to acid rain control within the United States. His quest for ways of "narrowing the political gap" separating the pro- and anti-control alliances is infused with a sense of political realism about how the American political system works and with a sense of the limits of political maneuverability available under prevailing conditions. As a consequence, he feels compelled to warn the reader that "as of mid-1985 in the acid rain issue the inertia of rest still appears much stronger than the inertia of movement."

His ultimate prognosis on the acid rain debate is that it will likely rage on without resolution for a number of years in the Congress, within EPA, and in the court system. If solution there be, he implies in his conclusion that the Swedish strategy of lake liming, in combination with some specific revisions in the present Clear Air Act and amendments, is probably the best "interim" program possible in the United States in the absence of some imaginative alternative yet to be fashioned.

11
The U.S. Politics of Acid Rain

George Clemon Freeman, Jr.

A CLASSIC POLITICAL STALEMATE

For over four years a stalemate has existed in Congress over proposed Clean Air Act Amendments. The cause of the stalemate is the impasse over acid rain legislation. That stalemate continues.

History of the Stalemate: The Players and Their Politics

The Senate

In August 1982, during the last days of the 97th Congress, the Senate Environment and Public Works Committee reported out a bill containing comprehensive Clean Air Act Amendments (S. 3041).[1] It included a modified version of Senator Mitchell's earlier acid rain bill,[2] which would have applied to the 31 "eastern" states,[3] imposed a 10 million ton rollback in sulfur dioxide (SO_2) emissions to 1980 levels and required "offsets" for emissions from new sources and increased emissions from existing sources (sometimes also called the "emissions growth CAP" or "the CAP"). The reported bill provided for a reduced 8 million ton SO_2 rollback in the same 31-state region, with certain exemptions for coal conversions. The emissions growth CAP prohibited all increases in SO_2 emissions above 1980 levels, including those authorized by EPA in any revised State Implementation Plan, unless the source obtained an offset. Among industries, the principal burden of the rollbacks would have been borne by the electric utility industry. (For perspective, annual gross SO_2 emissions of electric utilities totalled about 15.7 million tons in 1982.[4]) The impacts of the bill would have been most severe in states where utilities burn local high-sulfur coals. In these midwestern states the rollback requirements would also have had substantial adverse "secondary" effects on coal producers and electric energy-intensive industries and workers employed by those industries. In contrast, the emissions growth offset requirement would have produced greater social costs than the rollback requirements in the southern, "growth" states of the eastern half of the Sunbelt.

Breaking a long-standing tradition of unanimity this Senate committee of Senators Randolph (D-W.Va.),[5] Baker (R-Tenn.),[6] Domenici (R-N.M.) and Murkowski (R-Alaska)[7] voiced strong reservations about its acid rain provisions. And Senator Symms (R-Idaho)) voted against reporting it.[8] The acid rain provisions were also strongly opposed by key midwestern and southern Senators-- including Senators Warner (R-Va.), Byrd (D-W.Va.), Johnston (D-La.), Ford (D-Ky.), and Lugar (R-Ind.)--who were not members of the Environment Committee but who were on the powerful Senate Energy Committee.[9] S. 3041 was never brought to the floor for debate in the 97th Congress.

In the 98th Congress, a renumbered S. 3041 was introduced in the first session by Senator Stafford (R-Vt.) as S. 768. But its introduction was delayed until March 20, 1983, while Stafford sought joint sponsors among the then 16 members of the Senate Environment Committee. He ended up with only 10: Senators Chafee, Mitchell, Bentsen, Durenberger, Moynihan, Baucus, Humphrey, Domenici and Simpson. Six Senators--Baker, Abdnor, Symms, Randolph, Burdick and Hart--refused to sponsor it. Thus Stafford's attempt to repeat Senator Muskie's successful strategy of keeping the committee united in the 1976-1977 Clean Air Act Amendments failed. Other Senate bills dealing with acid rain were also introduced in the first session of the 98th Congress, but in practical political terms they were largely "bargaining chits." They included:

S. 145, introduced by Senator Mitchell (D-Maine).[10] It would have applied to the 31-state region and provided for a 10 million ton SO_2 rollback, plus an "offset" requirement for new sources and increased emissions from existing sources. It did not provide for spreading the cost of the control program beyond those responsible for achieving the rollback (i.e., "cost-sharing").

S. 769, introduced by Senator Stafford (R-Vt.),[11] This bill also would have applied to the 31 state region, required a 12 million ton SO_2 rollback by 1998, imposed an emissions "offset" obligation, and added mandated compliance with "best available control technology" (BACT) for all sources by 2000. It too did not provide for "cost-sharing."

S. 454, introduced by Senator Byrd (D-W.Va.) provided only for accelerated research.[12]

S. 766, introduced by Senator Randolph (D-WVa.)

called for a cap on SO_2 emissions, with exemptions for coal conversions, research and development grants for fluidized bed and other innovative technologies, and accelerated research.
S. 2001, introduced by Senator Durenberger (R-Minn.) would have required a 10 million ton SO_2 reduction in the 31 state ARMS region.[14] But the costs of these reductions would have been partially shared through funds obtained from a tax on SO_2 and NO_x emissions throughout the United States. The responsibility for reducing emissions would have been allocated by EPA through rulemaking rather than by the states through state quotas determined by a statutory formula. Offsets would have been required for new sources of SO_2 in all 50 states.
In the second session of the 98th Congress there were further developments. Senator Glenn (D-Ohio), then a serious contender for the 1984 Democratic presidential nomination, introduced S. 2215 on January 26, 1984.[15] His bill would have required an 8 million ton reducation in the 31-state ARMS region. The capital costs and 50% of the operating and maintenance costs of control equipment to attain these mandated reductions would have been funded through a 3 mill/kWh tax on electricity generated by fossil fuel-fired power plants within the 31 state region. The bill was structured to encourage attainment of the mandated reductions through installation of scrubbers and other advanced control technologies or by imposing limitations on switching from higher sulfur local coals to lower sulfur coals.
In November, 1983 and in February, 1984 the Senate Environment and Public Works Committee held hearings on S. 768. After two days of markups, the Senate Committee voted 16-2 on March 13, 1984 to report the bill out of committee.[16] During the 1984 markup, the committee adopted an amendment by Senator Humphrey (R-N.H.), as amended by Senator Mitchell, to increase S. 768's 8 million tons annual sulfur emissions reduction requirement to 10 million tons, and to require compliance by January 1, 1994, rather than January 1, 1995. The amendment also changed the bill's allocation "floor" from emission rates of 1.5 lbs/MBtu back to 1.2 lbs/MBtu.[17] The amendment also added an interim reduction requirement stipulating that compliance through fuel switching must be achieved within 5 years after enactment.[18] The bill contained no cost-sharing measures. Although reported out of committee, S. 768 never was brought to the Senate floor.
In the new 99th Congress, three bills intended to address acid rain have already been introduced in the Senate:

S. 52, introduced by Senator Stafford (R.Vt.). The bill is patterned after the acid rain portions of his bill from last year (S. 768), which contained comprehensive amendments to the Clean Air Act. The bill calls for a 10 million ton per year reduction in SO_2 emissions below actual 1980 levels, from the 31 eastern states by 1994. The bill would require offsets and it does not contain any cost-sharing provisions. The bill would also provide $5 million per year for three years to fund studies relating to the restoration of acidified waters.

S. 283, introduced by Senator Mitchell (D-Maine). This bill would also require a 10 million ton per year reduction in SO_2 emissions below actual 1980 levels, from the 31 eastern states within ten years after enactment of the statute. The bill would require offsets, and it does not contain cost-sharing provisions. It would, however, permit substitution of NO_x emission reductions for SO_2 reductions on a two-for-one basis.

S. 503, introduced by Senators Proxmire (D-Wisc.) and Humphrey (R-N.H.). This bill would also require a 10 million ton per year reduction below actual 1980 emissions. The bill would mandate the reduction of emissions in a two-phased program: 5 million tons by 1991 and 5 million additional tons by 1998, unless EPA requires less of a reduction in the second phase based upon the results of accelerated research during the first phase. During the first phase, accelerated acid rain research would be conducted. Then, prior to the second phase, EPA would evaluate the research results to determine whether any mid-course corrections should be made. The bill does not require offsets, but any coal conversion must meet a 1.2 lbs/MBtu emission limit. This bill, however, is modeled after the National Governors Association acid rain control recommendations.To the dismay of midwestern governors, however, the bill does not reflect any concession to the cost-sharing principle they were at least able to have recognized as meriting consideration in the National Governors Association's final resolutions.

As of the spring of 1985, these three bills were apparently near the bottom of the agenda of the Senate's Environment and Public Works Committee. Superfund reauthorization and the Clean Water Act Amendments have higher priority. It appears unlikely, therefore, that the Senate will act on acid rain legislation in 1985.

The House

In both the 97th and 98th Congresses, the House

Committee on Energy and Commerce was unable to report out any Clean Air Act Amendments. Acid rain was only one of several key Clean Air Act issues sharply dividing that committee in both Congresses. At the beginning of the first session of the 98th Congress two bills were introduced reflecting the two poles of political sentiment in the House:

H.R. 132, introduced by Representative Gregg (R-N.H.) called for a 10 million ton SO_2 emissions rollback in 48 states, with different requirements for East and West. It included no cost-sharing provision.[19]

H.R. 1405, introduced by Representative Rahall (D-W.Va.) was the same as Senator Byrd's bill. It provided for accelerated study only.[20]

Subsequently, on June 23, 1983, Representative Sikorski (D-Minn.), with the support of Representative Waxman (D-Calif.), introduced H.R. 3400--their proposed "compromise" on acid rain.[21] This bill would have required a 10 million ton SO_2 reduction in the 48 contiguous states. Scrubber retrofit would have been mandated at the 50 largest utility plant emitters, and partial "cost-sharing" of capital costs would have been funded by a 1 mill/kWh tax on all non-nuclear generation or importation of power. The bill would also have tightened NO_x limits for new trucks and new coal-fired powerplants.

On November 16, 1983, Representative D'Amours, on behalf of a coalition of representatives of northeastern states, introduced H.R. 4404, a somewhat altered version of H.R. 3400.[22] This bill would have required a 12 million ton SO_2 reduction by 1993 (11.2 million tons in the 31 eastern states and 800,000 tons in the West; 10 million tons from electric utilities and 2 million from industrial sources). It would have also provided a 1.5 mill/kWh tax on the sale of electricity from non-nuclear and non-hydroelectric generating stations (i.e. coal, oil and gas fired powerplants).[23] Representatives Dingell (D-Mich.), Broyhill (R-N.D.), Madigan (R-Ill.) and other key members of the full House committee never endorsed H.R. 3400 or H.R. 4404. Chairman Dingell remained strong in the view that any SO_2 program must apply nationwide and the NO_x emissions should not be further restricted.[24] In the 98th Congress' second session, despite a failure to negotiate a compromise with committee chairman Dingell, Rep. Waxman proceeded with consideration of the Waxman-Sikorski bill in his subcommittee, provoking the expected clash of regional interests.[25] The subcommittee did not resume acid rain discussions after a close May 2, 1984 vote (10 to 9) to reject the

Waxman-Sikorski bill. For all practical purposes, that ended consideration of acid rain legislation in the House in that session.

In the new 99th Congress, three bills have been introduced in the House:

H.R. 1030, introduced by Rep. Conte (R-Mass.) would reduce SO_2 emissions by 12 million tons per year below actual 1980 levels, within 10 years after enactment of the statute, and NO_x missions by 4 million tons below actual 1980 levels within 12 years after enactment. Conte offered this bill on behalf of New England Congressional Caucus; it would apply to the contiguous 48 states. The bill is similar to H.R. 4404, introduced by Rep. D'Amours in the 98th Congress. The bill would provide for partial cost-sharing through a 1.5 mill/kWh tax on electricity generated in the lower 48 states, except for hydroelectric and nuclear power. The bill calls for a two-phase program. First, a 6.5 million ton reduction in SO_2 would be achieved by requiring the 50 highest emitters to install scrubbers. In the second phase, the states would allocate the remaining 5.5 million ton reduction.

H.R. 1162 and 1414, were introduced by Rep. Green (R-N.Y.) and are essentially identical. These bills would reduce SO_2 emissions by 10 million tons per year below actual 1980 levels. The bill would apply to the lower 48 states and it would charge the states with the planning and implementation of emission reductions from existing stationary sources of sulfur dioxide. The bill would amend section 111 (New Source Performance Standards) to delete the percent reduction requirement for sources commencing construction after the bill is enacted and complying with 1.2 lbs/MBtu $SO2$ emission limit, and to revise NO_x emissions limits to 0.3 and 0.4 lbs/MBtu for sub-bituminous and bituminous coal (30-day rolling average), respectively.

The Reagan Administration

During the first two years of President Reagan's first term, his administration steadfastly opposed acid rain legislation. However, in announcing the nomination of William Ruckelshaus as EPA Administrator in May of 1983, the President indicated a willingness to reconsider the administration's policy on the acid rain issue. Ruckelshaus subsequently considered the issue at length, and there were reports that he was close to recommending a relatively limited emission control program. But these proposals encountered stiff opposition from OMB Director David Stockman and Secretary of Energy Donald Hodel. Thus in its reconsideration of this issue, the administration developed an internal stalemate comparable to that existing in both houses of Congress.[26] As a result, despite preliminary indications during the fall of 1983 that Ruckelshaus' return to EPA would presage a basic change in the administration's opposition to acid rain legislation, that was not the case.

President Reagan's subsequent 1984 State of the Union address adhered to the earlier "status quo." He did not advocate any tightening of controls, but instead announced additional funding for scientific research mitigation efforts and the development of advanced control technology.[27] In subsequent testimony before the Senate Committee on Environment and Public Works, Ruckelshaus confirmed that no legislative initiatives would be proposed in the near future. In support of the administration's "research only" stand, he cited the continuing uncertainty surrounding acid rain and its effects:

> On the basis of the current state of knowledge, the Administration is prepared to recommend additional sulfur controls.
>
> That does not mean the door is closed. It means that before launching the country an expensive and potentially divisive program, we feel we need more information.[28]

Ruckelshaus also stressed the administration's commitment to increase funding for acid deposition research.

Later in the year, when Administrator Ruckelshaus testified before the House Subcommittee on Health and the Environment of the Committee on Energy and Commerce, he identified four key scientific uncertainties:

> There are gaps in our knowledge in at least four areas that are relevant to forming sound acid rain policy. The first, and perhaps most basic, gap involves the scope of the problem. We really do not know the extent of the damage presently caused by acid deposition. . . . Second, we don't know at what pace this damage has occurred. . . . Third, we are uncertain as to whether or to what extent present levels of damage may be getting worse as a result of current levels of emissions. . . . Fourth, and finally, there are gaps in our knowledge of where the acid deposition in any given area comes from. This is the source-receptor problem. . . . These uncertainties and gaps in our knowledge

are matters of far more than academic interest. They are directly relevant to responsible public policy choices.[29]

The administration's commitment to increased research funds resulted in an increase in the planned budget of the National Acid Precipitation Assessment Program ("NAPAP"), the federal comprehensive acid rain research program from $27.6 million in fiscal year 1984 to $55.5 million in 1985.[30]

Under the administration proposal, the 1985 budget for the federal National Acid Precipitation Assessment Program would double and the program would focus on issues such as acidification trends, source/receptor relationships, monitoring of acidic levels in wet and dry deposition, effects on forests and aquatic bodies and the long term/short term acidification process generally. For fiscal year 1985, the administration also committed an additional $67 million program to develop new control technology and a $5 million demonstration project evaluating the feasibility of mitigating effects from acid deposition.

For fiscal year 1986, President Reagan's budget request for NAPAP is $85.5 million.[31]

The increased funding would launch research designed to (1) describe the sensitivity of aquatic resources to acidification and the rate at which acidification occurs, (2) establish the extent and severity of alleged forest declines in the United States, and (3) accelerate the installation of a monitoring network for dry deposition.

On January 5, 1985, Administrator Ruckelshaus left EPA and was replaced by Lee Thomas.[32] Since his confirmation, Administrator Thomas has remained loyal to the administration's position of supporting acid rain research rather than legislative proposals for new emissions reductions.

In March of 1985, the acid rain issue was one of the topics discussed by President Reagan and Canadian Prime Minister Brian Mulroney, at their St. Patrick's Day "Shamrock Summit." After that meeting, they announced the appointment of two special envoys: the U.S. envoy, Andrew Lewis; the Canadian envoy, William Davis. These envoys were given the following tasks: (a) to pursue consultation on laws and regulations that bear on pollutants thought to be linked to acid rain; (b) to enhance cooperation in research efforts, including that for clean fuel technology and smelter controls; (c) to pursue means to increase exchange of relevant scientific information; and, (d) to identify efforts to improve the U.S. and Canadian environment.[33]

Among the States

In 1983 a Task Force of the National Governors Association headed by Governor Sununu of New Hampshire was given the initial task of formulating a compromise on the acid rain issue. But that initial effort failed.[34] Earlier the Southern Governors Association had adopted a resolution opposing any acid rain proposals that included (1) an emissions "CAP" or "growth offset" requirement or (2) cost-sharing. Several weeks later, the midwestern governors defeated a resolution calling for a 50% SO_2 rollback proposal. Then in December of 1983 the northeastern governors endorsed a 10 million ton SO_2 emission reduction plan which also called for cost-sharing.[35]

Finally, in February of 1984 the National Governors Association (NGA) adopted a resolution calling for a two-step 10 million ton reduction program as well as a simultaneous $100 million nationwide research program.[36] The first phase of the control program would require a 5 million ton reduction in annual SO_2 emissions by 1990 in the 31 state ARMS region from 1980 levels. After a three year evaluation period, the second phase would require an additional 5 million ton reduction to be achieved by 1997. The EPA administrator would have discretion to revise or completely void this second phase reduction requirement. The NGA's recommendation side-stepped endorsing any explicit cost-sharing mechanism by only recommending that the costs be "shared by the states in a fair and equitable manner," and that credit be given for past investments to reduce emissions.[37] But by evading any specific solution to the cost-sharing dilemma, the NGA proposal did not provide a solution to legislative deadlock. The lack of a real consensus was apparent at the NGA's 1985 annual meeting, where Governors John Sununu (R-N.H.) and Anthony Earl (D-Wisc.) tried in vain to gain NGA approval of S. 503, the Proxmire-Humphrey bill that contains most of the NGA principles save one--cost-sharing. Governor Richard Celeste (D-Ohio) and other midwestern governors have severely criticized the bill for not reflecting NGA's cost-sharing policy.[38]

Soon after the NGA meeting, the Coalition of Northeastern Governors issued a new version of their acid rain control policy. It calls for a 10 million ton reduction of SO_2 emissions and a 4 million ton reduction of NO_x by 1995 in a two-stage program. The New England Coalition proposed a change to the funding provisions of its 1984 acid rain reducation policy by replacing the policy's combination emissions and kilowatt-hour tax with a scaled emission tax based on state-wide SO_2 emission rates. The tax which would apply to fossil-fuel generated domestic and imported electricity. The funds generated would be held for distribution to states in proportion to their share of total emission reduction requirements.[39] The coalition also proposed an interest-free loan program to finance acid rain controls on electric utility powerplants.[40]

Within Industry

After years of searching, the electric utility industry was unable to come up with a "consensus" on which particular SO_2 rollback proposals would be "least objectionable" if legislation were to be enacted or on whom should bear some or all of the costs of any further SO_2 or NO_x reductions. After months of their own internal reviews, both the Edison Electric Institute, the trade association of the investor owned segment of that industry, and the National Association of Rural Electric Cooperatives both strongly reiterated the case for "study only."[41]

The American Public Power Association (APPA) ultimately came out with a "cost-sharing" proposal for any mandated SO_2 or NO_x reductions. Under APPA's proposal, additional capital costs would be funded by a nationwide tax based on gross SO_2 and NO_x emissions.[42] This proposal was later incorporated in S. 2001, the acid rain bill introduced by Senator Durenberger. Finally, the directors of the Tennessee Valley Authority (TVA), the nations's largest governmentally owned power producer, endorsed acid rain SO_2 emissions reduction proposals.[44]

Like the electric utility industry, where the stakes are quite high, the coal industry has remained divided between high-sulfur and low-sulfur coal producers.[45]

Reasons for the Stalemate

Acid deposition is a complex <u>scientific</u> problem. It involves a number of key questions, such as:

What are the effects of acid deposition on soil, surface water pH, aquatic biology, crops, materials and human health? What is the nature, magnitude and extent of changes in resources? What are the possible causes of any changes? Which changes can be regarded as acid deposition-induced damage? What are the dose-response relationships between acid deposition and these damages? Where acidification effects are observed, what are the relative contributions of acid deposition and other sources of acidity? What part do land use practices play in producing the observed effects? What are the relative contributions of SO_2 and NO_x emissions to gross acid deposition? What are their relative contributions to acid deposition in any particular sensitive area or region? What are the relative contributions of natural and "man-made" sources to gross SO_2 and NO_x emissions? What are the relative contributions of different "man-made" sources of SO_2 and NO_x within the United States to total U.S. "man-made" emissions? What are the relative contributions of "wet deposition" and "dry deposition" to gross acid deposition? What are the relative contributions of local sources and distant sources to acid deposition in any particular area or region (the "local sources versus "long-range transport" issue)? What are the trends in long-term emissions of SO_2 and NO_x from "man-made" sources within the United States? What would be the average pH of "clean" rain in the absence of man's contributions? Will reductions in SO_2 emissions produce equivalent reductions in gross sulfate deposition? Will reductions in SO_2 emissions produce equivalent (or even predictable) reductions in acidic deposition in any particular area or region (the "linearity versus nonlinearity issues)? What effect on gross nitrate deposition and ozone levels would reductions in NOx emissions produce?

Acid deposition is also a complicated political problem (see Figure 11.1). For example:

Acidic deposition has pitted Canada against the United States.[46] This environmental problem has pitted region against region within the United States. It has pitted stationary sources against mobile sources. The acid rain issue has pitted utility sources against other industrial sources. This environmental controversy has divided the coal industry. Acid deposition has worsened the split between union and non-union miners and within the United Mine Workers driven a wedge between northern (high-sulfur) and southern (low-sulfur) miners. The acid rain debate has divided the utility industry. And, most important, acid deposition has divided both major parties.[47]

The proponents of acid rain legislation, principally the environmentalists and a number of northeastern and California legislators, have apparently convinced themselves and a number of others that acid rain is a problem that requires further immediate governmental action. But they have not yet come up with a politically saleable proposal for a "consensus" solution. The opponents of acid rain legislation, principally the electric utilities, a substantial part of the coal industry, a number of midwestern and southern legislators and the administration, have apparently convinced themselves and a number of others that the claims of damage to

FIGURE 11.1

Regional Issues

Canada
- Claims Pollution from South
- Acid Rain-Sensitive Areas
- Canada's Air Stds.
- Surplus Energy

Northeast
- Downwind
- Low Unit Emissions
- High Energy Costs
- Acid Rain-Sensitive Areas

Mid-West
- High Unit Emissions
- Lower Energy Costs
- Native Coal
- Acid Rain-No Sensitive Areas
- Depressed Economy

Other Issues
- Coal Industry(Hi vs.Low S)
- The Science
- Distant vs. Local Sources
- The West(Why include us?)

resources resulting from acid deposition are too
speculative, that the relationship between specific
emission reductions and specific deposition reductions
is uncertain, that the costs of present proposals for
substantial SO_2 emissions rollbacks and a cap on
increased SO_2 or NO_x emissions are unacceptably high,
and that current proposals would discriminate unfairly
against the midwestern and southern states. But they
have not yet come up with a politically saleable
proposal for a "consensus" solution that would put the
issue behind us.[48]

In recent years it has become increasingly
difficult to get any "non-consensus" legislation
through Congress in the absence some overriding
political necessity. Witness the fate in the last
several Congresses of proposed legislation on (1)
natural gas deregulation, (2) regulatory reform, (3)
fair allocation of antitrust liability among multiple
defendants, and (4) products liability reform. The
only major "non-consensus" legislation enacted has
involved areas where Congress had to act: taxes,
budget or foreign affairs. The main reason for the
necessity of consensus is institutional--the way
Congress is organized and how it works. In the Senate,
a determined minority of senators can usually block
legislation through a filibuster and other delaying
tactics. In the House, the Speaker and the rules
committee can usually prevent legislation which they
oppose from coming to the floor, and the powers of
committee chairmen can be used to delay and delay. And
finally, rarely is a presidential veto overridden.

KEY ISSUES FOR ACHIEVING CONSENSUS ON "THE MEANS" IN
ANY ACID RAIN LEGISLATION

If additional emissions controls are to be
imposed, two key policy areas where there must be
agreement are: (1) Who must do what, where, and when?
(the "Controls Strategy Issues") and (2) Who will pay
for it? When? And how? (the "Cost-Sharing Issues").
Let us analyze these two key sets of issues briefly in
the light of presently available scientific and
economic evidence and see what conclusions logic and
practical politics might support.

The "Controls Strategy Issues"

In trying to answer these control strategy
questions, we need to uncover the real objective of any
new legislation: Is the real objective to reduce gross
emissions of SO_2 and NO_x as an end in and of itself?
This might be characterized as "the emissions reduction

imperative," Or: Is the real objective to reduce specifically identified adverse environmental effects (aquatic life, forests, etc.) of acid deposition from man-made sources? This might be characterized as "the acid deposition objective."

A gross emissions reductions goal would parallel the "zero discharges" and "prevention of significant deterioration" goals of the present Clean Water and Clean Air Acts. Such legislative goals presuppose that significant environmental benefits will result from implementing the policy and those benefits will always be greater than, or at least commensurate with, the social costs of implementation or else that any balancing of social costs and benefits is irrelevant because of the overriding, subjective, legislatively-imposed, environmental imperative.

As to the likelihood or unlikelihood that any significant social benefits would result from any further mandated SO_2 emissions reductions, the best scientific evidence presently available suggests, first of all, that a case cannot be made at this time for appreciable health benefits or net benefits to crops. Second, it indicates that the recent forest damage phenomenon in higher altitudes of Europe and this country presently appears to be equally or more closely associated with NO_x emissions and ozone. On the other hand, the evidence lends support to the view that some benefits to aquatic ecosystems may result from further SO_2 reductions. But, such benefits would be expected to occur only in geographic regions susceptible to soil and water acidification, e.g., those in which soils in the watersheds have a low concentration of the alkaline substances that neutralize natural and man-made acid deposition. In the U.S., those areas are the Adirondacks, northern New England, the higher elevations of the Appalachians, the Sierra Nevadas, western Colorado, northern Wisconsin, Minnesota and western Michigan. In Canada, it is the vast area covered by the Laurentian Shield.

What kinds of reductions of emissions of what pollutants in what regions are likely to produce measurable aquatic benefits in these sensitive areas? The honest answer is: we really do not know at this time. Granting this uncertainty, what are the best scientific guesses currently available? To get the answer, let us concentrate on the evidence that has been made available to Congress over the past several years, starting with the evidence presented at hearings before the Senate Environment and the Senate Energy Committees,[49] the 1983 National Academy of Sciences report on acid rain,[50] and the NAPAP 1983 Annual Report to the Congress and President which only became available in mid-1984..[51]

In the 1982 Senate hearings, the majority of scientists testified that (1) there was sufficient evidence to support claims that man-made sources of SO_2 and NO_x were adversely affecting aquatic life in the sensitive areas of New York and New England but that (2) there was insufficient evidence of injury to crops, trees or other vegetation.[52] These scientists were divided over whether the legislative proposals then before the Senate would produce any appreciable--or even detectable--improvements in water quality in those sensitive areas.[53] Skepticism concerning whether appreciable "aquatic" benefits would result from the additional SO_2 emissions controls proposed was based on three points. The first is the recurring question as to whether further reduction in SO_2 and NO_x emissions would produce equivalent reductions in sulfate and nitrate deposition (the "linearity v. nonlinearity" issue).[54] The second is the evidence that "local" emissions, such as SO_2, NO_x, hydrocarbons and/or catalytic elements associated with oil combustion, may influence acid deposition levels in the sensitive areas as much as--or more than--"long-range" transport of SO_2 and sulfates (the "local v. long-range sources" issue).[55] The third is the fear that some of the SO_2 rollbacks might be environmentally counterproductive by causing increases in downwind ozone levels.[56]

The 1983 National Academy report cast further light on two of the major scientific issues in the acid precipitation debate. Concerning the "linearity v. nonlinearity" issue, the report concluded that "there is no evidence for a strong nonlinearity in the relationships between long-term average emissions and deposition."[57] This conclusion was cited by proponents of present SO_2 rollback legislation as supporting not only the need for acid rain legislation but also the details of then pending proposals. But that is not so.[58] For the report went on to point out that its conclusion on the linearity issue was limited by two other conclusions. First, the linearity conclusion applies only to eastern North America taken as a single region.[59] Second, "it can be stated as a rule of thumb that the farther a source is from a given receptor site, the smaller its influence on that site will be per unit of mass emitted" (p. 10). The report then acknowledged that "the ultimate strategy for dealing with acid deposition will depend on the application of realistic, validated models" and that "current models or analyses" are unreliable. Therefore, the report concluded, "[w]e cannot objectively predict the consequences for deposition in ecologically sensitive areas of changing the spatial pattern of emissions in eastern North America, such as by reducing in one area by a larger percentage than in other areas" (p. 11).

In June 1984 the Interagency Task Force on Acid Precipitation, the group overseeing the National Acid Precipitation Assessment Program (NAPAP), published its annual report on the results of its 1983 research. In addition to describing the present and planned research projects, the report also summarized "recent advances and uncertainties in the scientific understanding of the acid deposition phenomenon and its effects."[60] The report on the evolving science pointed out the substantial uncertainties that continue to exist regarding the identification and extent of possible effects caused by acid deposition and the relationships between emissions and deposition, and deposition and possible effects.

The <u>control strategy conclusions</u> that flow from all these basic "findings" depend on whether the basic legislative objective is: (1) <u>to achieve gross emission reductions of SO_2 and sulfates as ends in and of themselves</u> or (2) <u>to mitigate the impacts of sulfate depositions</u> ("acid rain") <u>in sensitive areas.</u> If the real, but as yet not publicly acknowledged, legislative objective is gross emissions reductions, then an across-the-board percentage reduction in SO_2 emissions from all sources--regardless of their current emission rates--would follow from the conclusions of the 1983 National Academy report.[61] But that is certainly not what any of the acid rain bills provide. If, on the other hand, the basic legislative objective is to mitigate the results of acid rain in sensitive areas in order to obtain aquatic benefits or forest damage mitigation, the efficacy of the SO_2 rollback provisions in the past and presently pending bills is unclear in the light of the National Academy report's findings.

As the report expressly cautions, its basic conclusion[62] does <u>not</u> mean that equivalent reductions will be obtained equally throughout that vast 31-state region. Indeed, as the report makes clear, the bulk of <u>dry</u> deposition is likely to be near its individual source and the impact of reductions at individual emission sources on <u>wet</u> deposition is quite attenuated beyond 600 miles (1000 kilometers). Finally, the report itself has been characterized as "flawed in several areas" by reviewers from the national laboratories.[63] Researchers representing Oak Ridge National Laboratory, for example, indicated that "[t]he conclusion of no evidence of nonlinearity is not borne out by the data."[64]

The Cost-Sharing Issues

One of the most persistent Senate critics of the presently pending acid rain bills, Senator Lugar (R-Ind.), emphasized this relationship in commenting on the National Academy report's policy implications in a subsequent opinion column of the New York Times:

> The widely publicized conclusion of the Academy report is that reductions in sulfur and nitrogen oxides will bring about corresponding reductions in acid rain. Less well reported, however, are two other conclusions. . . .
>
> The first. . . is that existing models of the effects of acid rain are scientifically uncertain. It is stated in the report that 'we cannot judge the consequences of emissions from such a region as the Midwest, for deposition in another region, such as the Adirondacks or southern Ontario.' The second conclusion is that the closer a source of emission is to an affected area, the greater the benefits that can be achieved from a given reduction in emissions.
>
> The report is thus as much an indictment of proposed acid rain legislation as it is a call for action.

Senator Lugar then went ahead to address the "cost-sharing" issue: "[T]he cost to the Midwest of the approach advocated in these bills would be staggering. Although precise increases in electric bills are hard to forecast, estimates have ranged from 7 percent to 31 percent for utilities in Indiana."

In addition to these severe customer cuts, Lugar noted that "the bills now before the Senate would force between 36,000 and 45,000 Midwestern and Appalachian coal miners out of work by encouraging utilities to switch from high-sulfur Midwestern coal to either low-sulfur Midwestern coal to low-sulfur coal from the West or to alternate fuels." In his view, "these costs are intolerable. And, given the conclusions of the Academy's report, they are also unnecessary."

What did Senator Lugar conclude from these facts? "We must move," he argued, "toward a broad, cost-sharing formula that involves all states responsible for the problem, while also encouraging research into cost-effective means of reducing oxide emissions."[65]

POSSIBLE WAYS OF OBTAINING A POLITICAL CONSENSUS

In my opinion, in the absence some new, innovative, fair proposal (yet to be unveiled), no political consensus on a mandated further SO_2 and/or

NO_x emissions reduction plan is probable unless there is also a consensus solution to the "cost-sharing" issues. In the first place, the direct costs of almost all proposals to date for attaining substantial annual SO_2 emissions reductions are substantial. The indirect costs in the form of increased unemployment in the coal industry are also substantial. Moreover, the reductions mandated would vary substantially from region to region, state to state, and from utility system to utility system within states. Thus, if not subsidized in whole or in part through one or more various "cost-sharing" proposals, the costs of these reductions would fall with great disparity on different groups of utility consumers and could severely impact employment in the high sulfur coal fields.

If emissions reductions designed to obtain annual SO_2 reductions in excess of 2 to 3 million tons are imposed on stationary sources in 8 or 10 states with relatively high emission rates,[66] the direct and indirect costs would substantially impact industrial and electric utility consumers in the Midwest and Southeast, if restricted to the 31 eastern states.[67] If scrubber retrofit, usually the least cost-effective means of compliance, were mandated as the only means of achieving reductions in order to avoid adverse impacts on high sulfur coal producers and their employees, electric utility consumers would be hit even harder.

If, on the other hand, the emissions reductions were imposed on sources in the form of a uniform percentage reduction requirement or on the basis of proximity to sensitive areas, either of these alternative strategies would severely impact industrial and electric utility consumers in states that for the most part already have quite stringent emission limitations on large stationary sources. In fact, despite their stringency, these states may still have fairly high SO_2 and NO_x emissions densities, measured in tons per year per square mile, largely because of urban concentrations of mobile NO_x emissions and non-utility SO_2 emissions. Thus, selection of any one of these three alternative control strategies, as the sole or principal means of obtaining the desired SO_2 emissions reductions, without providing for some form of "cost-sharing," would likely engender the opposition of enough senators and representatives to block passage of any legislation.[68]

Recognition of this political fact of life was manifested in 1983 when Senator Danforth (R-Mo.) withdrew as a co-sponsor of the Mitchell bill and instead recommended legislation to fund some of the costs of the mandated reductions through a uniform per kilowatt-hour surcharge on fossil-fired electric power generation through the 31-state region. Early in 1984,

with the New Hampshire primary ahead of him, Senator
Glenn also picked up on this idea as a compromise.
Later, the sponsors of H.R. 3400 proposed a 1 mill/kWh
fee on the generation or importation of electricity
produced by non-nuclear powerplants to fund the capital
costs of scrubber retrofit. But cost-sharing
apparently still has little political support outside
New England and California, since H.R. 1030 is the only
bill introduced thus far in 1985 that incorporates a
fee mechanism. But a 1 mill/kWh tax proposal, or a
variation of it, is unlikely to resolve the cost-
sharing dilemma for several reasons.

Such proposals would compensate only some capital
costs--i.e., largely scrubber retrofits. They would
not compensate for any of the substantial increased
operating and maintenance expenses of scrubbers or
compliance option of any additional costs under the
fuel-switching. Indeed, 1 mill/kWh fee would not even
be adequate to fund the capital cost of most of the
scrubber retrofits mandated.

There are, moreover, strenuous objections to
having the costs of a new gross SO_2 or NO_x emissions
reductions policy discriminatorily imposed upon only
some of the sources of SO_2 and NO_x emissions and
sources which emit no SO_2 and NO_x (e.g., hydroelectric
facilities). This objection was clearly reflected in
the stance of the American Public Power Association and
the subsequent 1984 Durenberger bill. Preliminary
analyses indicate that the cost borne by electric
utility ratepayers would be about three times greater
under a nationwide kilowatt-hour tax than they would be
if a tax were instead imposed on all fossil fuels
(coal, oil and natural gas).

Areas (such as much of the West) which at present
do not preceive acid deposition problems and have very
low levels of emissions may be loathe to help solve
what they perceive as other people's problems.[69] Even
in areas like New York and New England, where the
aquatic benefits are desired, some consumers might
begrudge increased tax payment to help achieve
reductions in emissions mandated for other regions
downwind of them. This feeling could be particularly
strong where electric rates are already substantially
higher as a result of currently more stringent SO_2
limitations and other factors. This is undoubtedly one
of the reasons that the National Governors Association
has consistently side-stepped explicit endorsement of
any particular cost-sharing plan in its acid rain
proposals and instead is on record as recommending that
the costs of reducing acid rain should be shared by the
states in a "fair and equitable manner." New England
representatives, however, have indicated that they are
clearly in favor of some form of nationwide cost-

FIGURE 11.2

Cost Sharing Issues

- Should emission reduction costs be shared?
- Shared by whom?
- To what extent?
  - Capital costs (full/partial)
  - Operating expense (full/partial)
- If cost sharing- what mechanism?
  - Tax on all energy
  - Tax on gasoline
  - Tax on all fossil fuels
  - Tax on KWH's
  - Tax on emissions
  - Tax credit concept
  - General treasury funding
  - Other forms

sharing through co-sponsorship of H.R. 4404 in 1984 and H.R. 1034 in 1985.

Narrowing the Political Gap

If political consensus on so controversial an issue as acid rain is to be reached, what approaches might be taken to reduce the existing political gap separating the positions of various participants in the public controversy? Certainly, one strategy might be to <u>hold down the total costs of acid rain control</u>. Costs, of course, are directly affected by a number of factors--including the size of any mandated reductions, the means by which such reductions may be achieved, and whether or not an emissions CAP is also imposed.[70] Another alternative would be to <u>spread the costs over a wider base</u>. Various alternatives have been suggested (see Figure 11.2) and include the U.S. Treasury or special taxes on either all fossil fuels or SO_2 or NO_x emissions. Some believe that any special U.S. tax would have to apply nationwide to be constitutional. Of all the "funding" proposals advanced so far, a "fuel tax," which would apply to all fossil fuels under a formula designed to avoid distortions in interfuels competition (for boiler fuels, space heating, etc.,) would be the least disruptive of existing market relationships. The fuels tax alternative--which would be burned by all sources of gross emissions of SO_2 and NO_x would give greater flexibility in fairly excepting particular classes of emitters from additional emissions reduction requirements (e.g., autos, oil home heating units, small industrial users, perhaps even smelters) since they would still bear some of the costs incurred by other sources not exempted from any further emissions reductions requirements.

Yet there are additional questions that must be answered in any further search for consensus. For example, where should additional SO_2 and NO_x emission limitations be imposed? The alternatives proposed so far include: the 48 contiguous states (H.R. 1030), the 31 eastern states (S. 52, S. 283, and S. 503), or some 4-13-odd midwestern and northeastern states containing or adjacent to "sensitive" areas (Ohio, West Virginia, Pennsylvania and New York or those states plus Maryland, Delaware, New Jersey and the New England states which former EPA Administrator Ruckelshaus was reported to have considered.)[71] What pollutants (or "precursors") should be further limited? SO_2 and NO_x (H.R. 1030)?[72] SO_2 only (S. 52)? Primary sulfate emissions and/or reactive hydrocarbons too? Which sources of pollutants should be further limited or controlled?[73] All sources of pollutants (implicit in the National Academy's report)? Only stationary sources? Only utility stationary sources (S. 52)? Or perhaps more mobile source restrictions (H.R. 1030)?

Still other issues come to mind. How are the amounts of any mandated reductions to be determined? Will a national emissions limitation ceiling on existing plants be required in an analogous way to current NSPS--e.g., 4, 3, 2, or 1.2 lbs/MBtu of SO_2 for fossil-fired plants?[74] Or should total tonnages be allocated to each state under a statutory formula, and in turn be reallocated by each state to individual systems or sources?[75] And, if tonnage reductions or percent emission reductions are mandated, will they be measured against "actual" or "allowable" emissions in the base year?[76] What are to be the deadlines for attaining any mandated reductions? Is there to be a one- or two-step target date scenario? Note that, if it is the latter, a deletion of certain emission reductions or at least mid-course corrections to the existing control strategy should be triggered before the second phase. No less important, what means of compliance are to be allowed or mandated? Any means of compliance (scrubbers, fuel switching or shut down)? Only scrubbers?[77] Can mandated emissions be obtained by statewide or system-wide or plant-wide "trade-offs" ("bubbling")?

To continue, are limitations in terms of permitted tons of emissions per year (a CAP) to be imposed on increases in emissions over 1980 emission levels? If so, will the CAP apply to existing plants on the basis of "allowable" or "actual" emissions in 1980? 1983? or the higher of 1980 or 1983? Will the CAP also apply to new plants meeting the NSPS and/or the BACT requirements? Will trade-offs be permitted in meeting any CAP restrictions? Plant or system bubbling? State or area or national bubbling? Purchases of "pollution rights"? Will exceptions be allowed either from mandated reductions ("rollbacks") or CAP requirements for coal conversions? space limitations? financial limitations? or emergencies? Alternatively, is lake liming to be encouraged--directly or indirectly?

Last, but not least, are other overtly political issues to be resolved. How is any legislation to be coordinated with the provisions in any proposed treaty or agreement with Canada?[78] And how is any legislation or treaty to be reconciled with national energy and economic policies?[79]

THE PROSPECTS FOR CONSENSUS

The Administration's Stance

During the second half of 1983, EPA Administrator

Ruckelshaus and the White House Cabinet on Natural Resources and Environment were never able to come up with "consensus" recommendations to the President. Alternative recommendations that were the center of controversy would have: (1) required some legislatively mandated rollback requirements by a target date, which would be a part of a "demonstration project experiment" (these reductions would be confined to a limited number of states centering on New York, Pennsylvania, Ohio and West Virginia and perhaps others adjacent to them); or (2) imposed a CAP on 1980 emissions in additional states. Some form of cost-sharing of at least some capital costs (perhaps as "R&D") was apparently discussed. These proposals drew strong opposition on scientific, economic and political grounds. Thereafter President Reagan reaffirmed his administration's position in his 1984 State of the Union address. He did not advocate any tightening of controls, but instead made a commitment to double funding for scientific research and to fund mitigation efforts and development of advance control technology.

The latest stirrings within the Administration have been in the international arena--negotiations with Canada. What may or may not evolve from those efforts remains to be seen. But, in those negotiations, the U.S. representative has far less maneuverability than his Canadian counterpart because of the lack of unanimity within the U.S. over further SO_2 reductions and the extreme difficulty that governmental subsidization of further reductions would encounter--problems that do not apparently confront the Canadians.

Straws in the Wind

The point is well taken that whether or not there is ever any legislation on "acid rain" may come down to one of three choices: (1) modest reductions without cost-sharing (no direct or indirect scrubber retrofit mandated)[80]; (2) large reductions with cost-sharing; or (3) a completely new approach, such as recasting the Clean Air Act into the mode of the Clean Water Act, by applying nationwide technology-based standards on existing plants, taking cost-benefit principles into account, and perhaps agreeing with Canada that both countries will meet each other's more stringent standards for both stationary and mobile sources of SO_2 and NO_x. In the ongoing debates, it will be also interesting to see if the environmental groups on Capitol Hill seek to revive their earlier de facto alliance of 1977 with high sulfur coal interests in support of legislative dictation of technological means of compliance with any emission reduction requirements.[81]

Finally, in considering the likelihood of legislation in either 1985 or 1986, two other points are relevant: Congress has other, higher priorities-- the reauthorization of Superfund, the Clean Water Act Amendments, and the review of the Safe Drinking Water Act and Hazardous Air Pollutants--and these are likely to prove controversial and time-consuming. In politics, like other aspects of life, inertia usually prevails. And as of mid-1985 on the acid rain issue the inertia of rest still appears much stronger than the inertia of movement,

A Shift to EPA and the Courts

With the prospects of Congressional action on acid rain growing dimmer each year, the real fight for imposition of additional reductions on gross emissions of SO_2 appears to be shifting to the administrative and litigative forums--EPA and the courts.

At EPA we see pressures being brought to bear for reductions in gross annual tonnages of SO_2 emissions roughly equivalent to those sought in the current pending acid rain bills through complex indirect means. These include pressures for (1) imposition of a new, one-hour ambient air quality standard for SO_2, (2) changes in air quality modeling requirements, (3) changes in enforcement policies for existing state implementation plans and new source performance standard requirements, (4) more stringent "tall stacks" regulations, and (5) a host of other new or potential changes in EPA regulations and policies.

The courts are also serving as a forum for the acid rain debate. In March 1984, a coalition of states (New York, Pennsylvania and Maine) and environmental groups filed suit against EPA for its failure to take certain actions allegedly required of the agency under the interstate (sec. 126) and international (sec. 115) provisions of the Clean Air Act. The suit alleged that EPA had failed to adjudicate pending interstate air pollution petitions within the time-frame required by section 126 and failed to address alleged acid rain impacts in Canada under section 115 of the Act. The U.S. District Court for the District of Columbia ordered EPA to issue a final decision on the section 126 petitions, which were originally filed by the states in 1980 and 1982. The petitions requested the EPA use section 126 to require emissions reductions from midwestern states that allegedly contributed to acid deposition in the Northeast. On December 10, 1984, the agency published in the Federal Register its denial of the section 126 petition. EPA concluded that the evidence before the agency did not demonstrate that sources in the named midwestern states are preventing

the attainment or maintenance of the other Clean Air
Act requirements in the petitioning states. Regarding
the section 115 portion of the suit, the court is
currently considering whether it has the power to
entertain the petition.

AN ULTIMATE OBSERVATION

It appears that the issue of acid deposition will
be with us for some time to come in three forums:
Congress, EPA and the courts. And the scientific and
political debate is likely to rage on for a number of
years, no matter what is or is not done legislatively,
administratively or judicially. Thus, it is ironic
that, among the many different proposals that have been
made at the national level for ameliorating the
perceived adverse effects of acid deposition on aquatic
ecosystems in sensitive areas, there is no major
political support as yet for following the example of
Sweden during the current European phase down in SO_2
emissions.

The Swedes believe that they have a much greater
acid deposition problem than we do and that theirs has
existed much longer than ours. They are now engaged in
a substantial lake liming program to mitigate the
immediate aquatic effects of acid deposition. It is
also relatively inexpensive.[82] Meanwhile they and
their European neighbors are cooperating in a gradual
30% phase down effort in gross SO_2 emissions while
accelerating research projects to find ultimate causes
and possible long term cures to the generic problems.
Moreover, lake liming may ultimately prove to be an
indispensable part of the best practical long-range
solution to lake acidity problems since present studies
indicate that such acidity is influenced far more by
the high acidity present in forest litter and wetlands
and by the poor buffering capacity of watershed soils
than by acid precipitation.[83]

Such an "interim" program in the U.S. could be
coupled with legislation assuring that credits be given
for SO_2 or NO_x reductions effected after a specified
date in any future mandatory acid rain legislation
reduction program, should one be enacted. This would
assure that uncertainties engendered by the current
legislative impasse would not discourage further
voluntary SO_2 or NO_x reductions and that full credit
would be given for reductions to meet other new EPA
regulations under the existing Clean Air Act.
Meanwhile under existing law, new plants will of course
have to meet new source performance standards, best
available control technology requirements and
applicable prevention of significant deterioration

increments. Repeal of the percentage removal requirement in section 111 of the present Clean Air Act, which governs new source performance standards for coal-fired electric generating plants, would also further the goals of cleaner air and reduction of acid precipitation by making the construction of new, low emitting plants much cheaper. This would eliminate the current economic premium the current law affords to the continued operation of older, inefficient, and higher emitting coal-fired and oil-fired generating facilities.[84]

Unfortunately, American politics has shown an alarming tendency over the past two decades to become more and more polarized on many issues along ideological lines. Adoption of such a practical interim approach, therefore, may be beyond the capacity of our present system. An alternate explanation may be that, despite their rhetoric, some of the key advocates of acid rain legislation and acid rain "regulation" are simply using it as a means to promote an ideological goal of mandating further massive reductions in gross SO_2 emissions as an end, in and of itself--regardless of whether the resulting social benefits will or will not justify the heavy social costs that they would impose on the American people.

NOTES

I wish to thank Peter S. Everett, Esquire, for his assistance in the preparation of an earlier version of this paper and Charles H. Knauss, Esquire, for his assistance in the current paper. While I have represented portions of the electric utility industry in connection with the Clean Air Act, the views expressed herein are my own.

1. S. Rep. No. 666, 97th Cong., 2d Sess. (November 15, 1982) (hereinafter cited as "S. Rep. on S. 3041").
2. S. 1706, 97th Cong. 1st Sess. (introduced October 6, 1981).
3. In the vernacular of acid rain bills, the "eastern" United States consists of the 31 states east of and bordering on the Mississippi River. This area is also sometimes called the "ARMS" or "ARMES" region (for Acid Rain Mitigation [Environmental] Strategy).
4. EPA, National Air Pollution Emission Estimates--1940 to 1982, EPA Document No. EPA-450/4-83-024 (February 1984). The document expresses the tons in "teregrams" of 14.3.
5. Add. views of Sen. Randolph, S. Rep. on S. 3041, pp. 124-125.

6. Add. views of Sen. Baker, S. Rep. on S. 3041, p. 104.
7. Senator Domenici and Senator Murkowski's reservations were expressed verbally at the mark-up. Senator Murkowski reiterated them in his separate statement in the committee report. See **Acid Precipitation and the Use of Fossil Fuels: Hearings before the Senate Committee on Energy and Natural Resources**, 97th Cong., 2d Sess. 110 (August 19, 1982) (hereinafter referred to as "**Senate Energy Committee Hearings**") pp. 110-111. A year later, however, Sen. Domenici indicated support for the 1982 Senate bill's acid rain proposals. See **Inside EPA**, Vol. 4 (December 2, 1983), p. 7.
8. S. Rep. on 3041, p. 99.
9. See, e.g., **Senate Energy Committee Hearings**, pp. 11-13, 18-19, 109, 112, and 154-158.
10. He ultimately gained 16 co-sponsors: Senators Baucus (D-Mont.), Cohen (R-Maine), Cranston (D-Calif.), Dodd (D-Conn.), Hart (D-Colo.), Hollings (D-S.C.), Humphrey (R-N.H.), Kasten (R-Wis.), Kennedy (D-Mass.), Leahy (D-Vt.), Moynihan (D-N.Y.), Pell (D-R.I.), Proxmire (D-Wis.), Rudman (R-N.H.), Stafford (R-Vt.) and Tsongas (D-Mass.).
11. Co-sponsored by Senators Cranston (D-Calif.), Hart (D-Colo.), Durenberger (R-Minn.) and Humphrey (R-N.H.).
12. Ultimately it was co-sponsored by 15 Senators: Andrews (R-N.D.), Denton (R-Ala.), Dixon (D-Ill.), Ford (D-Ky.), Heflin (D-Ala.), Huddleston (D-Ky.), Inouye (D-Hawaii), Lugar (R-Ind.), Mattingly (R-Ga.), Melcher (D-Mont.), Percy (R-Ill.), Quayle (R-Ind.), Symms (R-Idaho), Trible (R-Va.) and Warner (R-Va.).
13. And co-sponsored by Senator Specter (R-Penn.).
14. 129 Cong. Rec. S14557 (1983).
15. 130 Cong. Rec. S237 (1984).
16. Senators Randolph (D-W.Va.) and Symms (R-Idaho) voted against reporting the bill.
17. The original Mitchell bill, S. 1706, 97th Cong. 1st Sess. (1981), had provided for the 1.2 lbs/MBtu floor but this had been increased by the committee to 1.5 lbs in the reported S. 3041 on a motion by Senator Moynihan, presumably to facilitate then pending New York utility oil to coal conversions that would otherwise have been forestalled, imposing substantial costs on New York consumers.
18. The provision was a significant departure from the "freedom of choice" in compliance principle that its sponsors and western supporters had proclaimed earlier. By accelerating the deadline for compliance for utilities that might prefer to lower emissions by

using lower sulfur coals, this provision could have substantially escalated short-term demands beyond immediately available supplies and thus forced low-sulfur coal prices up substantially. This may explain in part why some western senators from states with low-sulfur coals supported it. Such a price rise in low-sulfur coal prices could have in turn tilted more utility decisions toward the "high-sulfur coals using scrubbers" alternative. The result thus could have been higher electricity prices across-the-board to the public and U.S. industry.

19. H.R. 132, 98th Cong., 1st Sess. (1983).
20. See 129 Cong. Rec. H 512 (1983).
21. Waxman and Sikorski ultimately garnered about 130 co-sponsors, principally representatives from the Northeast and California, and most of them Democrats.
22. Its 22 co-sponsors were drawn entirely from the Northeast and were mostly Democrats.
23. Since New England and New York have agreed to purchase substantial amounts of power from Quebec's giant hydro-electric capacity over the next several years, exemption of hydro would result in a lesser impact on electric consumers in those states. Giving that additional competition advantage to Canadian hydrogenerated power in U.S. bulk power supply markets has been sharply attacked. The funds collected by this tax would finance the installations of .scrubbers, development of advanced control equipment, and mitigation measures.
24. See e.g., **Inside EPA**, Vol. 4 (September 30, 1983),p. 11; Vol. 5 (March 23, 1984), p. 7; Vol. 5 (March 16, 1984), p. 9.
25. See **Inside EPA**, Vol. 5 (March 2, 1984) p. 3; **Environment Reporter**, Vol. 14 (March. 23, 1984) ("Regional Divisions on Acid Rain Reflected in House Panel Testimony on Air Act Bills"), pp. 2083-2084.
26. EPA Administrator Ruckelshaus subsequently approached the White House with a proposal to establish an SO_2 emission control program, under section 115 of the Clean Air Act. See **Inside EPA**, Vol. 45 (January 6, 1984). Under section 115, EPA may require emission reductions it deems necessary to prevent or eliminate "air pollution which may reasonably be anticipated to endanger health or welfare in a foreign country." The attempt to invoke section 115 encountered resistance at the White House sufficient to derail the plan. The agency's consideration of section 115 proved ironic, in view of the fact that EPA was later sued for its alleged failure to comply with the provision.
27. See **Washington Post**, January 26, 1984, p. A-17.
28. **Acid Rain, 1984, Hearings before the Committee on Environmental and Public Works**, 98th

Cong., 2nd Sess. (February 2, 7, 9 and 10, 1984), p. 9 (hereinafter referred to as **Senate Acid Rain 1984 Hearings**).
    29. See **Statement of EPA Administrator Ruckelshaus before the House Subcommittee on Health and the Environment,** 98th Cong. 2nd Sess. (March 29, 1984), pp. 1-2.
    30. See **Senate Acid Rain 1984 Hearings,** p. 15.
    31. NAPAP, "FY 1986 Budget Summary" (February 27, 1985).
    32. For Ruckelshaus' departing views on acid rain, see **Environment Reporter,** Vol. 15 (January 4, 1985), p. 1452.
    33. U.S. State Department, "Press Release--The Quebec Summit: Joint Statement on the Environment" (March 17, 1985).
    34. See **Inside EPA,** Vol. 4 (October 28, 1983); and see earlier story **Inside EPA,** Vol. 4 (October 7, 1983), p. 125.
    35. See **Inside EPA,** Vol. 4 (October 7, 1983), p. 10.. See **Inside EPA,** Vol. 4 (December 2, 1983), pp. 1, 7-8. **Environment Reporter,** Vol. 14 (December 9, 1983), p. 1413. The resolution adopted calls for a two-phase plan, resulting in an estimated 10 million ton annual SO_2 emissions reduction and a 4 million ton NO_x emissions reduction by 1995. It would have applied to the 48 states.
    36. See **Inside EPA,** Vol. 5 (March 2, 1984), p. 7. The reduction requirement for each state within the 31-state ARMS region would be fixed for the first phase as equal to that state's relative share of electric utility SO_2 emissions at average rates in excess of 1.2 lbs/MBtu. EPA could alter this 1.2 lbs/MBtu formula during the second phase. The western governors, in states where the nation's smelters are located, were thus able to avoid inclusion of these major sources of SO_2 emissions in the NGA plan. This was a major political triumph for them since smelters are reported to account for 80% of SO_2 emissions west of the Continental Divide. The southern governors also won their fight against an overall emissions CAP on SO_2 emissions based on 1980 emissions. No cap on new SO_2 or NO_x emissions would be imposed other than the imposition of the current New Source Performance Standards (NSPS) for new coal conversions (i.e., 1.2 lbs/MBtu), and a requirement that EPA promulgate a small boiler NSPS. The states would otherwise be free to design their individual control strategies, engage in interstate trading of emission reductions, and substitute reductions in NO_x emission or SO_2 on a 2 for 1 basis. Finally, the governors' proposal called for an agreement for similar reductions to be negotiated with Canada. It is probable that the

governors did not realize the irony of setting a precedent for exempting smelters from any acid rain treaty with Canada.

37. In contrast, the $100 million research program would be federally funded.

38. See **Environment Reporter**, Vol. 15 (Marach 1,1985), p. 1814. According to Celeste, this bill does not reflect the position of the NGA because the association had rejected the "polluter-should-pay" philosophy.

39. See **Environment Reporter**, Vol. 15 (March 8, 1985), p. 1849.

40. See **The Energy Daily**, Vol. 13 (March 5, 1985), p. 3.

41. See **The Edison Electric Institute Position on Acid Control Legislation** (August 15, 1983); **Inside EPA**, Vol. 4 (September 2, 1983), p. 141; and statement of William H. Megonnell, Director, Legislative Affairs for Environment, Edison Electric Institute in **Senate Acid Rain 1983 Hearings** (November 17, 1983), pp. 330-331, 422-434. See, also, statement of Rae Cronmiller on behalf of the National Rural Electric Cooperative Association, FF **ibid.**, pp. 332-335, 435-455.

42. See **The Recommendations of the American Public Power Association (APPA) on Acid Rain Control Legislation** (August 1983) and **APPA's Policy Resolution on Acid Rain Control Legislation** (approved by membership, May 3, 1983).

43. The APPA believed that the Durenberger bill was "conceptually sympathetic" to the electric utility industry. But APPA disagrees with the bill on specific points: the amount of reductions called for are thought too high and too costly and the reductions are not adequately phased over time; also, APPA strongly opposes an emissions CAP.

44. See testimony of S. David Freeman, Richard Freeman and Charles H. Dean, Jr., **Senate Acid Rain 1984 Hearings**, pp. 237-259.

45. See **The National Coal Association Proposals for Revising the Clean Air Act and Regulations**, (December 8, 1980) and the **Acid Rain Policy Statement of the Alliance for Clean Energy**, (November 1983).

46. See **Environment Reporter**, Vol. 15 (February 22, 1985), p. 1767; Vol. 15 (March 15, 1985), p. 1913; Vol. 15 (March 22, 1985), p. 2012. See, e.g., **Inside EPA**, Vol. 6 (February 15, 1985), p. 5; Vol. 5 (March 16, 1984), p. 10; Vol. 4 (December 16, 1983), pp.4-5; (October 21,1983), pp. 1, 4, 13; (September 30, 1983), p.13; (September 16, 1983), p. 2. New Canadian Prime Minister Brian Mulroney has recently succeeded in having the environment ministers of Canada's federal government and several eastern provinces agree to reduce by 1994 the SO_2 emissions by 50% below the

actual levels of 1980--**Inside EPA**, Vol. 6 (February 15, 1985), p. 5. The goal of the program is to reduce 1980 SO_2 emissions by 50% or (approximately 1.9 million tons) to 2.3 million tons by 1994. The provincial governments will provide financial assistance to those required to reduce emissions except for the smelting industry.

47. For example, in the 98th Congress, 8 Senate Republicans took public positions urging enactment of acid rain rollback and emissions CAP provisions, while 12 took positions that reflect opposition to such proposals. Among Senate Democrats, 12 senators appeared to favor such legislation and 8 to oppose it. In the House, 72 Democrats and 9 Republicans sponsored it. In the House, 72 Democrats and 9 Republicans sponsored H.R. 3400 (Waxman-Sikorski) or its New England counterpart, H.R. 4404; while 40 Democrats and 33 Republicans sponsored its nemesis, H.R. 1405 (the Rahall Bill). On the "emissions growth offset" issue, for example, the electoral vote of the 9 Southern states that would be adversely impacted by such a provision in certain then pending Senate bills was 97 votes. If the offset requirement were extended nationwide, the electoral vote of the remaining Sunbelt states adversely affected was 95 votes. Turning to the high-sulfur coal-mandated scrubbers versus low-sulfur coal-free compliance choice issue, the electoral votes of states with predominantly high- sulfur coal reserves was 125 votes, those with predominantly low-sulfur coal reserves 59 votes, states with mixed reserves having 42 electoral votes. The electoral vote split over the cost-sharing issue was also sharply divided, with the Northeast and Midwest (and perhaps California) with about 300 electoral votes pitted against the rest of the West and the South, which together have about 200 electoral votes. Making a separate analysis of states within the 31 "eastern" state region, the electoral votes of states with substantial support for SO_2 rollback totalled 120 (New England, New York, Minnesota, Wisconsin and Michigan). The remaining midwestern states and the southern states where opposition against rollback requirements appears strong had 189 electoral votes. If the acid rain issue is still around in 1987 and 1988, the strong regional division on potential solutions will likely pose similar dilemmas for potential presidential candidates in both major parties.

48. Senator Stafford in December of 1983 said that Congress "still hasn't found the key to breaking down the regional differences that are so much a part of this problem." See **Environment Reporter**, Vol. 14 (December 9, 1983), p. 1412. And little has changed since 1983 to supply the missing key. Indeed, the

battle lines still seem firmly drawn over six key issues: (1) 31 versus 48 or 50 states, (2) zero versus a 5 versus an 8 versus a 10 versus a 12 million ton rollback in annual SO_2 emissions from 1980 levels, (3) mandated scrubbers versus freedom of choice in compliance with any mandated rollbacks (e.g., switching from high to lower sulfur coals), (4) an SO_2 emissions growth cap versus no cap, (5) cost-sharing versus no cost-sharing and (6) further NO_x versus SO_2 controls (mobile versus stationary sources).

49. See **Senate Environment Committee Acid Rain hearings** (May 25 and 27, 1982); **Senate Energy Committee Hearings** (August 12, 1982); **Senate Acid Rain 1983 Hearings** (August 25, October 14, and November 2, 15, 17, 1984); **Senate Acid Rain 1984 Hearings** (February 2, 7, 9 and 10, 1984).

50. The National Research Council, **Acid Deposition: Atmospheric Processes in Eastern North America** (Washington, D.C.: National Academy Press, June 1983) (Hereinafter referred to as **"The National Academy Report"**).

51. The Interagency Task Group on Acid Precipitation, **1983 Annual Report on the National Acid Precipitation Assessment Program to the Congress and the President** (Washington, D.C.: U.S.G.P.O., June 1984).

52. See, e.g., **Senate Environment Committee Acid Rain Hearings**, pp. 225 (Dr. Cowling), 235 (Dr. Evans), 237 (Dr. Arthur Johnson), and 238 (Dr. Dale Johnson).

53. Even the scientists who felt contemporaneous legislation could be justified could not agree on what form it should take. See, e.g., **Senate Environment Committee Acid Rain Hearings**, pp. 227 (Dr. Cowling), 240 (Dr. Loucks), 13 & 87 (Dr. Holmes), and 87-88 (Drs. Shannon and Oppenheimer).

54. **Senate Environment Committee Acid Rain Hearings**, pp. 6, 7, 13, 14-15, 59-60, 82, and 83-84. For a recent report furnishing persuasive support for nonlinearity see: Dr. Perry Samson, "On the Linearity of Sulfur Dioxide to Sulfate Conversion in Regional-Scale Models" [64-66], Report to the Office of Technology Assessment (June 1982).

55. The evidence presented at both the May 25 and 27, 1982, hearings in support of one or more of the local source theories was substantial. Nine witnesses testified before the Senate Environment and Public Works Committee on May 25, 1982. Seven of them gave testimony that provides some support for one or more of the "local sources" theories. See, e,g., **Senate Environment Committee Acid Rain Hearings**, pp. 6-7, 100 (Dr. Demerjian); 86, 124-125 (Mr. Hicks); 12-13, 126-130 (Dr. Homes); 15-16, 23, 55-57 (Dr. Mohnen); 21-22, 55, 77, 213-214 (Dr. Rahn); 24-25, 87 (Dr. Shannon); 67

(Dr. Barrie). In addition to the evidence cited at the Senate Environmental Committee, see also the extensive Report on "Acid Rain: Commentary on Controversial Issues and Observations in the Role of Fuel Burning," prepared for the U. S. Dept. of Energy's Morgantown Engergy Technology Center for PEDCO Environmental, Inc. (the "PEDCO Report"), Febraury 1982, which is included in the **Senate Energy Committee Hearings**, pp. 261-472.
    56. See, e.g., **Senate Acid Rain 1983 Hearings,** p. 16 (Dr. Mohnen).
    57. **The National Academy Report**, p. 7.
    58. The irony of this misuse of the National Academy report to call for additional controls on SO_2 emissions for stationary sources but not on NO_x was heightened by the report's extraordinary reliance on the Hubbard Brook data. See, e.q., **The National Academy Report**, p. 140. These data show (1) relatively level or declining rates in annual sulfate deposition, but (2) increase peaks in short term, episodic **nitrate** deposition. Indeed, Dr. Noye Johnson, a principal in the ongoing Hubbard Brook study, indicated in his Senate testimony in 1982 that the increasing "peaks" in acid deposition of concern at Hubbard Brook were "more directly related to nitric acid and internal combustion engines, than to sulfuric acid." **Senate Environment Committee Acid Rain Hearings**, p. 244.
    59. **The National Academy Report**, pp. 7-11.
    60. Interagency Task Force on Acid Precipitation, **1983 Annual Report on National Acid Precipitation Assessment Program to the Congress and President** p. 13 (Washington, D.C.: U.S.G.P.O., June 1984), p. 13.
    61. Subsequent attempts to claim health and visibility benefits for SO_2 reductions by environmental groups supporting acid rain legislation provide indirect evidence that gross emissions reductions **per se** may be their true legislative objective. See, **e.g.**, the statement of Richard Ayers for the Clean Air Coalition, **Senate Acid Rain 1984 Hearings**, pp. 509-510.
    62. The report concludes that the same percentage reduction in **annual** SO_2 emissions from **all** SO_2 emission sources in the **entire** eastern region will produce a roughly equivalent reduction in the percentage of sulfate deposition, both dry and wet, within that region **taken as a whole.**
    63. Letter from Dr. Bernard Manowitz, Chairman, National Laboratory consortium of the Interagency Task Force on Acid Precipitation to the Honorable Dan Fuqua, **et al.**, Vol. 1 (Aug. 30, 1983). Dr. Manowitz further expanded the National Laboratories view at the Senate Environment and Public Works Committee hearing of November 2, 1983--**Senate Acid Rain 1983 Hearings**, pp. 395-396.

64. Letter from Dr. Bernard Manowitz, Chairman, National Laboratory consortium of the Interagency Task Force on Acid Precipitation to the Honorable Dan Fuqua, **et al.**, Oak Ridge National Laboratory comments p. 3. See, also, the House of Representatives press release by Rep. Lloyd (D-Tennessee) and Rep. Dingell (D-Michigan), dated March 9, 1984, which quote a statement approved by Dr. Jack Calvert and Dr. Bernard Manowitz clarifying the relationship between the Calvert committee N.A.S. report and the national laboratory's critique.

65. **New York Times**, August 15, 1983.

66. Or a reduction of only several million tons if concentrated in only two or three states, as was advanced by administrator Ruckelshaus in the fall of 1983.

67. If, however, acid rain legislation applied nationwide and applied to gross SO_2 emissions from all major sources, consumers in western states, where smelters contribute far more than utilities to gross SO_2 emissions, could also be severely impacted.

68. The equity issue posed by "cost-sharing" cannot be answered by simply saying that the polluter should pay. Who is the polluter? (A) Anyone who emits any pollutant, notwithstanding his current emission rate, with whatever emission control measures (technological or fuels content) he has elected? or (B) Only those who now (before any further amendment) are in excess of current Clean Air Act requirements? Note that the latter is how "pollution" is defined under the Clean Water Act. Most of the electric utilities that would be affected by acid rain SO_2 rollback requirements in the pending bills are now in compliance with **all** present Clean Air Act requirements, and thus are now in full compliance with the law. Only after enactment of any acid rain legislation would they become "polluters" to the extent that they violated any **new** requirements. The present wide variations throughout the country in legally specified emission rates for SO_2 from powerplants and other industrial sources are more attributable to historic and economic factors than to state governmental choices as to what is required to meet all presently applicable Clean Air Act requirements or whatever higher standards of purity a state may elect.

69. Tangible evidence of substantial mountain states political opposition was the letter of 15 senators to President Reagan, dated November 17, 1983, that urged "an equitable and cost-effective solution" to the acid rain problem but at the same time strongly opposed cost-sharing. The letter also opposed forced scrubbing as a means of compliance. See **Inside EPA**, Vol. 4 (December 2, 1983), p. 1.

70. I again emphasize that if tonnages are substantially reduced but concentrated in only a few states, while **total** national costs may be reduced, the impact on the "chosen" states may be equally severe or greater once indirect costs such as adverse competitive effects and unemployment increases are taken into account. It should also be noted that the substantial costs of an emissions CAP on states where new growth is otherwise probable and states with low levels of economic activity in 1980 appear to have been overlooked by many in recent public and private discussions of acid rain alternatives.

71. For an example of the argument made for starting with such a more limited area in imposing additional controls, see Goklany, "Near and Distant Area's Contribution to Acid Deposition in the Northeastern U.S.," **The Environmental Forum**, Vol. 2 (November 1983), p. 20. His thesis is that despite their shortcomings present SO_2 models are still the best way to estimate the likely results of "acid rain" reduction strategies.

72. The decision to concentrate on SO_2 emissions and to ignore NO_x emissions in most of the bills that have been introduced to date probably reflects a recognition of the extreme political difficulties that would confront efforts to impose additional restrictions on mobile sources. It may also reflect a recognition of the practical difficulties inherent in securing additional reductions at stationary sources. While some burner modifications may achieve reductions at relatively modest costs, add-on NO_x removal systems are in their technical infancy. Man-made NO_x emissions, however, are now suspected of being the principal cause of acid deposition concerns of Hubbard Brook. It is also of interest that the recent report by Environmental Resources, Limited, takes the position that in Europe additional controls on NO_x emissions merit as much or greater attention as additional SO_2 controls--See, Environmental Resources Limited, "Acid Rain, A Review of the Phenomenon in the EEC and Europe (London: Graham and Trotman, Ltd., 1983) [hereinafter referred to as **The EEC Report**].

73. See **The EEC Report**, pp. 18-19 [which considers only three means of additional SO_2 controls: (1) a requirement for use of flue-gas desulfurization in all **new** coal-fired powerplants, (2) further desulfurization of oil fuels and (3) substitution of low sulfur imported coals].

74. Another approach to an SO_2 (and NO_x) emissions reduction program for existing sources would be a variation on the Clean Water Act's provisions for "best practicable control technology currently available" requirements for pollution control of

effluent discharges from "existing" plants. Such Clean Air Act BPT could allow compliance by any practicable means, such as clean fuels and "front end" technologies as well as scrubbers. EPA could set such requirements from classes or categories of emission sources, such as coal-fired powerplants, smelters, industrial boilers, oil-fired home heating units, etc., taking into consideration age of plant, economics, potential adverse environmental effects of alternatives, likelihood of benefits, etc., coupled with a meaningful variance procedure.

75. If we are going to talk in terms of state emissions, "emissions in tons per square mile" of inhabited territory of both pollutants (SO_2 and NO_x) should be the relevant measure of each state's relative contribution to national combined gross emissions. It is misleading to talk as if total annual tonnages on a statewide basis were the relevant figures in assessing a state's relative contribution to the acid deposition problem. Total annual tonnages of emissions are irrelevant to the fairness of any proposed reductions. Only if we talk about emissions in terms of tons per square mile can we avoid distortions caused by the vast and fortuitous differences in the relative sizes of our states.

76. In 1984 Canada announced a 30% reduction target against "allowable" rather than actual emissions. See Keating, "Canada is Committed to Spending Billions on Acid Rain Clean Up," **Globe and Mail Reporter** (March 8, 1984). Recent U.S. bills, on the other hand, speak in terms of reductions from actual emissions in the base year 1980. In the United States, actual utility emissions in 1980 were substantially below allowable emissions on most systems. Thus Canada's 30% target of 1983 would have been only about 17% if viewed in comparable terms of the then pending U.S. legislation. More recently, however, Prime Minister Mulroney announced a new target of 50 percent reduction from actual emissions in 1980. **Inside EPA**, Vol. 6 (February 15, 1985), p. 5.

77. See the 1984 bill, H.R. 3400, Section 185(c)(1)(A), or the 1985 bill, H.R. 1030.

78. An interesting question posed by any U.S.-Canadian negotiations is whether both countries could agree to the imposition of the same emission limitations on the same classes of SO_2 and NO_x emission sources and on a common cost-subsidization policy. If not, how can they otherwise avoid substantial discrimination among competing U.S. and Canadian producers in the same industries?

79. The need for such a reconciliation preoccupied Congress' consideration of the 1977 Clean Air Act Amendments in both 1976 and 1977. But this

subject has received virtually no attention to date in Congress' consideration of acid rain proposals.

80. Two to three million tons if spread over the 31-state region, or a proportionately lesser amount if confined to only three or four states.

81. See, e.g., Ackerman and Hassler, **Clean Coal/Dirty Air** (New Haven, Conn.: Yale University Press 1981), pp. 54-55; Haskell, **The Politics of Clean Air, EPA Standards for Coal-Fired Generating Plants** (New York: Frederick A. Praeger, 1982), pp. 12 and 16; Peter Navarro, "The Politics of Air Pollution, **Public Interest**, Vol. 36 (Spring 1980), pp. 39, 43 and 59.

82. **The EEC Report**, p. 142.

83. See, e.g., A.G. Everett, W. C. Retzsch, J. R. Kramer, I. P. Montanex, and P. F. Duhaime, **Hydro Geochemical Characteristics of Adirondack Waters Influenced by Terrestrial Environments** (Rockville, Md.: Everett & Associates, Decmber 1982); A. V. Molliter and D. J. Daynal, "Acid Precipitation and Ionic Movements in Adirondack Forests," **Soils Science Society Journal**, Vol. 46, (1982), pp. 137-141; **The Integrated Lake-Watershed Acidification Study: Proceeding of the ILWAS Annual Review Conference**, prepared by Tetra Tech, Inc., for the Electric Power Research Institute, EA-2827, January 1983.

84. Current acid rain SO_2 reduction proposals will only achieve a short-term (20 plus years) acceleration in the projected long-term decline in utility gross annual SO_2 emissions likely to result from the replacement of older plants with new ones meeting New Source Performance Standards. The rate at which gross SO_2 emissions form utilities is currently uncertain because current new source performance standards and "prevention of significant deterioration" substantive and procedural requirements provide deterrents against the retirement of older higher emitting coal-fired plants at the end of their currently projected lives. But there are legal as well as practical limits to indefinite use of these units. EPA's reconstruction and replacement criteria (40 C.F.R. S 60.15) at some point would trigger compliance with New Source Performance Standards.

Appendix

Acid Rain and the Social Sciences: A Selected Bibliography

Elizabeth W. Yanarella
and Ernest J. Yanarella

INTRODUCTION

In recent years, the primary and secondary literature on the acid rain phenomenon and its effects has grown immensely. The following bibliography offers annotations of 27 important books, articles, monographs, reports, and special journal issues on a wide range of key issues and topics featured in the acid rain debate. Studies from the physical and biological sciences, economics, law, journalism, political science, and policy analysis are represented in this annotated listing in order to provide some sense of the breadth of the mounting corpus of writing in this area and to demonstrate the relevance of diverse fields of scholarly inquiry to the transcendence of the policy stalemate over acid rain control. Its purpose will be fulfilled if it serves as a useful introductory guide for the budding acid rain researcher in the social and policy sciences to a number of the most essential materials in this new and exciting field of scholarly endeavor.

Ackerman, Bruce A., and William T. Hassler. **Clean Coal/Dirty Air.** New Haven, Conn.: Yale University Press, 1981.

This scholarly study explores the consequences of the demise of the independent, administrative agency responsible for regulating problems without being captured by special interest groups. Using the Clean Air Act as an example of an agency facing statute, it shows how hyperbolic symbols, congressional overload, regional interests, and a lack of full inquiry led to a statute which ignored the very ecological concerns and economic principles it was designed to protect and which instead protects a special group--in this case, the high-sulfur coal industry.

Ackerman and Hassler support environmental bills which set an end of compliance and allow the polluter the freedom to determine the means of achievement. The level of compliance, they feel, should be established only after a full inquiry into: the nature of the problems being confronted, available and forseeable technological fixes, and the economic feasibility and enforceability of its institution. While some policy analysts have claimed that the authors have overstated their argument, the book remains essential background reading for an understanding of some of the administrative and policy conundrums produced by the Clean Air Act which will have to be dealt with in developing legislative means for responding to the ecological problems of acid precipitation.

Boyle, Robert H., and R. Alexander Boyle. **Acid Rain.** New York: Nick Lyons Books, 1983.

This portrait of the various aspects of and theories concerning acid rain and its environmental consequences originally appeared as a series of articles in **Sports Illustrated.** It is primarily aimed at the lay reader and concerned citizen. Though clearly a veiled polemic against what is seen as a major environmental threat, the book lays out persuasive arguments relying heavily on journalistic observations and interviews and numerous scientific studies and experiments in the United States. Freely drawing upon personal hunches and inferences, the authors sometimes point to suspected harmful environmental and human health dangers from acid rain where scientific research to date is mute or ambiguous. One relatively new facet of the "death" of bodies of water highlighted in this book is the danger to rare aquatic insects whose place in the evolutionary chain has not yet been established. Perhaps its strongest chapter is the one exploring the tangle of regional interests and administrative stalling techniques contributing to the political stalemate over acid rain control policy in the United States and to the diplomatic immobilism caused largely by the United States in dealing with the U.S.-Canadian transboundary pollution problem.

Brown, Susan. "The International Law and Pollution Control and the Acid Rain Phenomenon." **Natural Resources Journal.** 21 (1981). 631-646.

In this wide-ranging article, the author takes a look at existing acid rain legislation, the indigenous political and economic obstacles to negotiating an international licensing procedure and standard for dealing with the problem, and the influence of the "international law of neighborliness" which increasingly compells governments to seek workable bilateral solutions. Although the article is oriented to the international facets of airborne pollution, the author does focus some attention on the content and objectives of the Clean Air Act and subsequent amendments, indicating those factors contributing to the ineffectiveness of this environmental legislation. Her solution to overcoming the act's deficiencies involves the administrative strengthening of the Environmental Protection Agency to allow for better monitoring and enforcement of SO_2 emissions standards and to deny state implementation plan relaxation.

Carroll, John E. **Environmental Diplomacy: An Examination and a Prospective of Canadian-U.S. Transboundary Environmental Relations.** Ann Arbor, Mich.: The University of Michigan Press, 1983.

An interdisciplinary look at the environmental/diplomatic interactions between the United States and Canada, this volume analyzes key transboundary environmental problems in North America, such as issues of offshore oil negotiation, adjacent wilderness-development areas, and air and water control and quality. Carroll develops a careful historical perspective on Canadian/American relations with special emphasis placed upon the differences in approaches and backgrounds of the two North American partners and the difficulties which have occurred in cases where a largely federal form of government undertakes to negotiate with an essentially provincial form.

The chapter on acid precipitation is largely a report on the findings of the scientific investigations generated by the joint U.S.-Canadian Research Consultative Group on the Long-Range Transport of Air Pollutants and challenges of those findings by industrial sources and more recently administrative elements in the United States. Current legislative proposals for alleviating the effects of regional and cross-boundary acid rain transport are discussed. As this inquiry amply illustrates, while key political and economic elites tend to view the problems associated with acid precipitation as predominantly domestic matters, they are in reality increasingly becoming transnational issues for bilateral negotiation.

Comptroller General of the United States, General Accounting Office. **An Analysis of Issues Concerning "Acid Rain."** [GAO/RCED-85-13] Washington, D.C.: U.S. General Accounting Office, December 11, 1984.

This report by the General Accounting Office is an

examination of a wide spectrum of issues fueling the acid rain debate--including the causes and effects of acid deposition, the alternative strategies for controlling acid deposition, and the potential economic impacts of such control strategies. Based on available scientific work, the study finds that scientific evidence alone does not lead unequivocally either to the conclusion that a control program should be undertaken immediately or to the conclusion that acid deposition control should await better understanding. Despite its fence-straddling position, the GAO survey is a valuable first source to social scientists interested in acquainting themselves with the technical and political-economic complexities of the acid rain problem.

Crocker, Thomas D., ed. **Economic Perspectives on Acid Deposition Control.** Stoneham, Mass.: Butterworth Press, 1984.

This collection of essays originated out of the Symposium on Acid Precipitation in conjunction with the American Chemical Society's annual conference held in Las Vegas in the spring of 1982. Each individually-authored chapter focuses upon some dimension of the acid deposition problem deemed to be useful to policymakers dealing with this or similar environmental issues--including the economic benefits of control, the extent to which control specification should be optimally carried out locally or across the entire polluting region, the various methods and strategies of control, the legal aspects of transboundary pollution, and the policymaker's delicate act of fashioning policy which balances the risks of excessive costs and irreparable ecological damage. Overall, the essays generate more questions than they answer and in the process point either to major scientific uncertainties or to the limitations of normative economics as obstacles to rational environmental policymaking.

Crocker, Thomas D., and James L. Regens. "Acid Deposition Control: A Benefit-Cost Analysis: It Prospects and Limits." **Environmental Science & Technology.** 19 (February 1985). 112-116.

America's continuing economic doldrums and the Reagan administration's budget-tightening approach to social welfare and environmental programs have recently elevated the role of benefit-cost analysis as an instrument of policy evaluation. In the acid rain debate, this instrument has been frequently employed in the political arena to assess various control strategies generated in the executive and legislative branches of the federal government. After reviewing the kind of information required for legitimate benefit-cost analysis and then surveying the benefits and costs of control which have been presented in past studies, Crocker and Regens underscore the hazards and limitations of this economic tool, concluding that "the value of benefit-cost analysis of acid deposition control resides more in its potential contributions to clearer statements of the problem than in its provision of accounts for social and ecological bookkeeping." In other words, they seem to be suggesting, neither scientific research nor economic benefit-cost accounting alone or in combination will supplant the necessity of political choice on issues in this or other policy arenas.

Electric Power Research Institute. "Acid Rain Research: A Special Report." **EPRI Journal.** 8 (November 1983). 1-54.

In a collection of cautiously-worded essays, this special issue enumerates the strategies and tentative findings of a series of high-focused research projects on acid precipitation supported by the Electric Power Research Institute. The general pattern of the arguments suggests that, while perturbations and changes in the ecosystem are evident, the sources of such environmental degradation cannot be conclusively proven to be the result of airborne pollution. The acknowledgment that "some examples of lake acidification may have been caused by man-made pollution" is perhaps the most strongly-worded statement regarding the relationship between ecological disturbances and socially-caused acid rain. Other conclusions on the effects of acid deposition upon forests, farmlands and crops, and human health are couched in the most hypothetical language and tend to be attributed to natural, rather than anthropogenic, causes. On the other hand, unlike the negative, obstructionist approach on acid rain policy taken by the Edison Electric Institute, EPRI has adopted a more responsible corporate strategy to influence the direction and substance of the acid rain debate by underwriting extensive scientific research aimed at acquiring a sound research foundation based upon modeling techniques. Whether such research will foster timely policy to ameliorate the effects of acid rain or will instead merely help to confine the policy dispute to narrow technical issues and scientific uncertainties which can never be entirely resolved is conjectural.

Gold, Peter S., ed. **Acid Rain: A Transjurisdictional Problem in Search of a Solution.** Buffalo, N.Y.: Canadian-American Center, State University of New York at Buffalo, 1981.

This edited transcript of the first major conference on the technical and social aspects of the acid rain phenomenon provides an excellent overview of competing positions taken on this issue among government policymakers, environmentalists, scholars, and utilities representatives. In the first section of the conference proceedings, noted scientists and environmental officials grapple with the nature and seriousness of the problem and the short- and long-term impact of acid deposition upon the environment. There, the scientific debate over acid rain is nicely highlighted by the presentation of contrasting views by Eville Gorham, the respected ecologist from the University of Minnesota, and Ralph Perhac, the director of environmental assessment at the Electric Power Research Institute. Other presentations examine the problem of acid rain in various Canadian provinces.

The other major section of the book addresses the policy question--i.e., should acid rain be reduced? Drawing upon the views and expertise of policy analysts with various institutional affiliations, these presentations investigate the economic/environmental tradeoffs which they believe might be involved in a sound acid rain control policy using different models of cost-benefit analysis. Again, the perspectives on this issue are carefully balanced between those--like William Poundstone and A. Joseph Dowd, representatives of the coal and electric power industries-- who urge great caution in estimating the benefits of control measures and those--like Thomas Crocker and Raymond Robinson, an American economist and Canadian environmental official--who suggest that conventional analytic techniques are likely to underestimate the benefits of acid rain controls and conceal the true costs of acid rain over the long term.

In addition to two concluding papers which focus on available policy alternatives for controlling acid rain and on public attitudes toward such control, the appendix contains several treaties and agreements concerning air and water quality which bear directly or indirectly upon acid rain diplomacy in North America and Europe.

Havar, Magda; Hutchinson, Thomas C.; and Gene Likens. "Red Herrings in Acid Rain Research." **Environmental Science & Technology.** 18 (1984). 176A-186A.

This essay by three acid rain researchers was written in an attempt to repudiate the five most common misconceptions in the acid rain controversy perpetuated by anti-control forces in industry, science, and the political realm. Having reached the conclusion that acid rain controls are necessary, the three noted scientists cite findings from an extensive body of research to disclaim explanations offered in popular writings which divert public attention from the significant and indisputable industrial and other man-made sources of acid rain and its harmful environmental effects. Based upon their reading of the research record, they argue that the time is now passed when such "smokescreens"--so strenuously generated by the same industries which produce smoke--should be allowed to hamper serious efforts to resolve these problems.

Howard, Ross, and Michael Perley. **Acid Rain: The Devastating Impact on North America.** New York: McGraw-Hill Book Company, 1980, 1982.

Summing up the main thrust of the book in the preface to its second edition, the authors characterize it as written to "reveal the virtual state of environmental emergency concerning acid rain and the pathetic, almost duplicitous political and corporate response to it." This environmentalist critique is roughly divided into two parts--the first half devoted to elucidating the known and suspected hazards and costs of acid rain (termed the "stepchid of pollution") with the second half addressed to the political techniques of polluters and other anti-control elements to undermine impending regulations. Both the footdragging approaching of the United States and the "you go first" approach of Canada are bemoaned by the authors, who conclude that neither tactic will yield an ultimate winner in the bilateral diplomatic game. The book is particularly enlightening in relating the election-year strategies of key U.S. Senators and President Carter to the original failure in 1979 and 1980 to control acid rain.

Johnston, Douglas M., and Peter Finckle. "Acid Precipitation in North America: Quest for Transboundary Cooperation." **Vanderbilt Journal of Transnational Law.** 14 (1981). 787-843.

This scholarly article is essential reading for anyone beginning to research into international law and the problem of pollution control. Notable both for its depth and breadth of presentation, the essay analyzes such issues as: anticipated harm and spatial dimensions (an interpretation of which is necessary to devise economically appropriate controls), dispersal and reduction methods, legislative strategies for promulgating ambient or emissions standards, transnational legal principles applicable to pollution control, and multilateral treaty arrangements extant in this policy realm. One very pertinent section is the legal analysis of the only area of tort law relevant to pollution control—nuisance—and the obstacles which plaintiffs have to overcome to receive damages.

Surveying the policy options, the authors reject the alternatives of only one country imposing stringent controls or of both Canada and the United States gradually phasing in less than stringent controls, believing that either would only mitigate the cumulative effect of damage from acid deposition. The only solution is the negotiation of a bilateral accord requiring common standards and compatible industrial controls which are economically appropriate to the anticipated harm. Moreover, they claim, new plants must install the best equipment (not just the least-cost alternative); existing plants must install the most advanced retrofit equipment; and small existing plants must install retrofit equipment as is economically feasible over the next twenty-five years.

Knudson, Duane A., and David G. Streets. "Mitigation of Acid Rain—Policy Alternatives." **Environmental Progress.** I (1982). 146-153.

This article examines several methods for reducing SO_2 emissions within the electric utility industry by using the Utility Simulation Model developed by Teknekron Research, Inc. The means of compliance are chosen by the utilities on a least cost basis with the only constant being, in two cases, the employment of local coal. While a thorough analysis is given, the authors conclude by arguing that state specific features determine the actual effectiveness of a national strategy. The states selected for modeling analysis were Ohio, West Virginia, and Illinois.

LaBastille, Anne. "Acid Rain—How Great a Menace?" **National Geographic.** 160 (1981). 651-681.

This lushly photographed, carefully-researched article in one of America's most popular nature magazines follows in the tradition of earlier finely-crafted articles in **National Geographic** focusing on controversial environmental issues in the United States and around the world. After tracing the history of acid rain from its natural origins many centuries ago as a soil fertilizer to its transformation into an "environmental time bomb" with the acceleration of the Industrial Revolution, LaBastille examines the full panoply of demonstrated and suspected environmental effects of acid precipitation. Suggesting that the hidden costs of acid rain may have already surpassed the expenses of controlling it, she urges a policy for mitigating acid rain impacts which combines energy conservation and emissions control technology.

While most useful for public education purposes, several of its charts should be of interest to social scientists.

Likens, Gene E., et al. "Acid Rain." **Scientific American.** 241 (1979). 43-51.

Despite the absence of long-term studies of acid rain in the United States, scientific research into the acid rain phenomenon has pointed to a lowering of the pH in rain in certain parts of America such that precipitation in some regions are five to thirty times more acidic than the normal pH of rain. In this classic scientific article, Likens and his associates provide an excellent overview of the problem which social scientists just beginning to explore the social dimensions to this environmental problem should read. Both natural and anthropogenic causes of acidic precipitation are discussed and the difficulties of trying to isolate and measure each source are noted. Data and trends from European and American studies in precipitation chemistry are disclosed and variations in the measurement of surface water are accounted for by differences in geological substrates. This balanced and measured analysis concludes with a cautious exploration of an acid rain control policy characterized by improved control technologies plus fuel conservation.

Luoma, Jon R. **Troubled Skies, Troubled Waters: The Story of Acid Rain.** New York: The Viking Press, 1984.

Written by a journalist with long experience in the reportage of environmental and conservation concerns, this book should be considered a valuable pre-primer for the budding acid rain researcher in the social sciences. Although not organized with self-contained chapters focusing on individual aspects of the problem as are other books in this list, Luoma's loosely structured account is surprisingly thought-provoking. The work intersperses reconstructions of field visits to locations where the effects of acid rain are being concretely felt with journalistic accounts of the process by which the damage is occurring. Other chapters present an historical perspective on the studies and findings of over a century of scientific research on acid precipitation, a non-polemical discussion of a decade of bickering between the United States and Canada over transboundary acid rain effects, and a review of the debate between environmentalists and industrial representatives in the United States with examples of the publicity promoted by each group.

National Research Council, National Academy of Sciences. **Acid Deposition: Atmospheric Processes in Eastern North America.** Washington, D.C.: National Academy Press, 1983.

Following up on its earlier 1981 study of acid deposition, this work presents an update and a review of current scientific understanding of the processes linking emissions to deposition and the issue of the linearity vs. non-linearity in relations between emissions and deposition. Its two major conclusions, poorly worded by the authors and badly distorted by the media and certain interested parties, are: "that the relationship between emissions and deposition in northeastern North America is substantially non-linear when averaged over a period of a year and over dimensions of the order of a million square kilometers [i.e., a large geographical region]"; and "that if the emissions of sulfur dioxide from all sources in this region were reduced by the same fraction, the result would be a corresponding fractional reduction in deposition."

Social scientists will find its summary of the state of scientific research on atmospheric processes extremely useful in gaining quick familiarity with key terms and issues in the acid rain debate--especially the source-receptor and the linearity issues. The treatment of this scientific review in the political process would also make an interesting case study of the use and abuse of science.

National Research Council, National Academy of Sciences. **Atmosphere-Biosphere Interactions: Toward a Better Understanding of the Consequences of Fossil Fuel Combustion.** Washington, D.C.: National Academy Press, 1981.

This report was one of the first major studies of the acid deposition phenomenon by a leading scientific research organization in the United States. Its primary focus is on the impact on living systems of atmospheric pollutants related to fossil fuel combustion for energy production. On the basis of its review of available field data in sensitive areas and the results of various controlled experiments undertaken by scientists, it reaches the conclusion that with respect to vulnerable freshwater ecosystems, "[i]t is desirable to have precipitation with pH values no lower than 4.6 to 4.7 throughout such areas....[And] in the most seriously affected areas (average precipitation pH of 4.1 to 4.2). this would mean a reduction of 50 percent in deposited hydrogen ions." Establishing a pattern for later misreadings of such technical reports on acid rain, this conclusion was extensively misinterpreted in the mass media and in other forums as calling for a 50 percent reduction in the precursors of acid deposition (including sulfur dioxide and oxides of nitrogen).

"The Politics of Acid Rain." **Alternatives: Perspectives on Society and Environment.** 11 (December 1983). 1-48.

This special issue offers a collection of varied essays concerning political and legal aspects of the acid rain problem and its abatement. Especially valuable is the article, "Life, Liberty, and the American Pursuit of Acid Rain," by Don Munton. This piece examines the difficulties inherent in any issue which places "politically and economically strong groups of industrialists against an unmobilized public both within and without national borders." The asymmetries between the American and Canadian perspectives on the causes and amelioration of transboundary acid rain pollution are probed by John Carroll, while the stalling techniques of Canadian corporations antagonistic to controls are delineated by Phil Weller. James Kraus' essay, "Legal Approaches to the Control of Acid Rain," makes the case that impediments to the resolution of the problem is not legal in origins but rather lies in the political strength of an industry which provides essential products. In another piece, "Acid Rain: The Implications for Energy Policy," Ralph Torrie uncovers the unusual alliance which has been forged between Canadian environmentalists and the nuclear lobby; as he demonstates, these strange bedfellows have teamed up in the acid rain dispute to pose the policy choice of "atoms vs. acid." Like Pehlke's article, he too presents the soft energy path as a clean and durable solution to the continuing pollution of the environment by sulfur and nitrogen emissions.

Postel, Sandra. **Air Pollution, Acid Rain and the Future of the Forests.** [Worldwatch Paper No. 58] Washington, D.C.: The Worldwatch Institute, 1984.

In this monograph, evidence of the stress of acid deposition upon our global ecosystem--especially the forests--is presented by drawing heavily from recent scientific studies. These studies suggest significant forest damage in Germany's Black Forests and in parts of Eastern Europe, as well as more ambiguous tree dieback in portions of the U.S. Northeast. As Postel notes, even the less developed countries comprising the Third World are seeing signs of acid accumulation in their soil which by implication may begin to show up in the trees in LDCs. While she does not attribute instances of forest dieback to acid deposition alone, she concurs with forest experts about the susceptibility of trees to natural stresses when

weakened by stresses from man-made pollution.
Damage to forest, soils, and lakes, she argues, is an additional cost--an externality--of fossil fuel combustion. In effect, society is subsidizing fossil-fuel generated electricity by excluding these costs in the price of elctricity; and this, in turn, induces overconsumption of energy. Correction of this market failure and protection of the ecosphere, in her view, entail the formulation of a control policy incorporating technologies of reduced emissions and more efficient use of energy. One of the few glaring weaknesses of this environmentalist perspective on the acid rain problem is its failure to take into account mounting evidence from the latest research into forest damage which implicates ozone alone or in combination with acid rain as key culprit.

Rhodes, Steven L. "Superfunding Acid Rain Controls: Who Will Bear the Costs?" **Environment**. 26 (1984). 25-32.

In the course of offering a political and economic reading of pending congressional bills related to acid rain control, the author investigates the strengths and weaknesses of the funding strategy of establishing a superfund. Although in general supporting this funding option, this political scientist raises a number of ethical questions regarding that alternative--including the fairness of obliging "victims" to pay the "villains" and the equity of requiring nonpolluting states to subsidize those states considered major polluters. Despite its rather focused character, this article is one of the best evaluations of a particular funding mechanism.

Rhodes, Steven L., and Paulette Middletown. "Public Pressures, Technical Options: The Complex Challenge of Controlling Acid Rain." **Environment**. 25 (1983). 6-38.

In this lengthy article, the influence of political pressures bearing on the environmental policymaking process--often without strong scientific support--is viewed with particular reference to the acid rain controversy. Less an exercise in policy advocacy, the authors are primarily concerned with showing how prior protective policies in regard to the environment--e.g., the debate over the risks of chlorfluorocarbons in the atmosphere and the Great Lakes Water Quality Agreement--were made in a context of vigorous public lobbying and incomplete scientific evidence.
In the case of acid precipitation, they encourage the reduction of SO_2 and NO_x emissions while acknowledging that the complexity of the problem makes the level and

control of such reductions difficult to establish. Any one control measure (such as "scrubbers") might not be sufficient, they caution, and the levying of immediate controls might institute a control strategy which from hindsight might prove inadequate and even counter-productive. They conclude that polluting industries cannot escape the imperative that "the polluter must pay" at least some of the costs while environmentalists must accept the truth that there are no simple solutions or quick fixes to the acid rain problem.

Rosencranz, Armin. "Th International Law and Politics of Acid Rain." **Denver Journal of International Law and Policy.** 10 (1981). 511-521.

Although principles of transboundary pollution legislation derive from the maxim, "Use your own property in such a manner as not to injure that of another," Rosencranz develops the argument that nations will control pollution only if and when it is in their national interest to do so and not because of any purported obligation under international law. From his perspective, bureaucrats and legislators are motivated for the most part by concerns to protect national autonomy and economic self-interest. Thus, prospects for transnational action will be improved, not from national legislation by one country or the other, but from information exchange among international organizations which may eventually form a consensus on the seriousness of the problem--thus creating a favorable climate of opinion wherein policymakers and concerned citizens may influence national agendas.

Streets, David; Knudson, Duane A.; and Jack D. Shannon. "Selected Strategies to Reduce Acidic Deposition in the United States." **Environmental Science & Technology.** 17 (October 1983). 474A-485A.

This economic study examines a variety of methods of control of sulfur emissions in terms of their regional impacts, cost-effectiveness, resultant electricity pricing consequences, economic disruptions, and regional coal market effects. Using the Advanced Statistical Trajectory Regional Air Pollution Model based on linear source-receptor relationships, comparisons are made of the effectiveness of the methods surveyed in southeast Canada and eastern United States. Treated to secondary analysis are the possible controls for NO_x and receptor mitigation programs.

Streets, David G., et al. "A Regional New Source Bubble Policy: Its Advantages Illustrated for the State of Illinois." APCA Journal. 34 (1983). 25-31.

Proceeding from the thesis that existing the New Source Performance Standards structured into the Clean Air Act impedes the retirement of older utility plants, this article examines the value of a new source bubble policy using criteria such as state-wide cost, local air quality impacts, and impacts upon state coal mining economies. It finds that for the case of Illinois there is a clear reduction in the cost of a buble policy when compared with compliance with existing regulations. Their overall finding is that, in addition to providing the advantage of least cost, the bubble policy opens up new markets for midwestern coals.

U.S. Congress, Office of Technology Assessment. **Acid Rain and Transported Air Pollutants: Implications for Public Policy.** [OTA-0-204] Washington, D.C.: Office of Technology Assessment, June 1984.

Adhering to its customary approach of careful evaluation of policy questions generated by modern science and technology without strong policy advocacy, the Office of Technology Assessment (OTA) assigned to a project staff the task of sifting through the most recent technical analyses of acid rain and presenting various policy alternatives for congressional consideration. Noting that the issue of transported pollutants "poses a special problem for policymakers," the report seeks to consider and assess ways in which the Congress can "address current damage, consider potential harm to recipients of transported air pollutants, yet not enforce an unnecessarily high cost on those who would be required to reduce pollutant emissions." While eschewing easy answers, it lays out "some carefully weighed estimates of costs, some carefully reasoned conclusions about the nature and extent of downwind damages and risks, and several policy options that merit consideration." As an example of dispassionate policy evaluation, this report ranks as a model--whatever conclusions one draws about its findings and judgments.

U.S. Environmental Protection Agency, Office of Research and Development. **The Acid Deposition Phenomenon and Its Effects: Critical Assessment Review Papers. Volume I (Atmospheric Sciences) and Volume II (Effects Sciences).** [EPA-600/8-83-016A and B] Washington, D.C.: U.S. EPA, May 1983.

These two volumes comprise a comprehensive review of the state of scientific knowledge of precursor emissions, pollutant transformation to acidic compounds, pollutant transport and deposition, and the measured and potential effects of acid deposition. Covering the breadth of

technical issues involved in the acid deposition
phenomenon and its effects, the document offers chapter
reviews of the leading scientific research and findings on
those issues. While the social scientist will discover
little in the way of research with a direct bearing on the
social, economic, and political dimensions of acid rain
debate, this review is useful not only as a guide to
scientific knowledge of acid deposition. It is also
interesting for the light it sheds on the function played
by scientific research within the Environmental
Protection Agency and the larger policy process in the
search for political consensus on this environmental
question.

Wetstone, Gregory S., and Armin Rosencranz. **Acid Rain in Europe and North America: National Responses to an International Problem.** Washington, D.C.: The Environmental Law Institute, 1983.

This scholarly examination of the acid rain problem
in Europe and North America, prepared by two of America's
most knowledgeable legal experts on environmental law,
lucidly describes the nature and severity of the acid rain
phenomenon, the individual and joint steps in law and
policy being taken by nations to mitigate its effects, and
the actions being mounted by international organizations
seeking multilateral solutions to a growing global
environmental problem. This comprehensive survey is
essential reading for the social scientist or policy
analyst interested in the comparative dimensions of the
acid rain debate and the different approaches being taken
by various countries to surmount the effects of domestic
and cross-frontier impacts of this environmental problem.

A Note on the Contributors

JOHN E. BLODGETT is a Senior Analyst in Environmental Policy with the Congressional Research Service of the Library of Congress. He received a B.A. and M.A. from Washington State University and an M.A. and A.B.D. from Case Institute of Technology. His major subject areas with the Congressional Research Service include environmental policy and economics, toxics, pesticides, and acid rain. He has authored or co-authored several reports on acid rain, focusing on policy issues surrounding control of acid rain precursors and their impact on long-term environmental policy.

JOHN E. CARROLL is Professor of Environmental Conservation at the University of New Hampshire and a Kellogg Foundation National Fellow. He is the author of numerous books and articles in international environmental diplomacy, acid rain, and Canada-U.S. relations. His works include **Environmental Diplomacy** (University of Michigan Press, 1983) and, with Ambassador Kenneth Curtis, **Canadian-American Relations: The Promise and the Challenge** (Lexington Books, D.C. Heath and Company, 1983).

GEORGE C. FREEMAN, JR., is a Member of the Virginia, District of Columbia, and Alabama Bars. He received his L.L.B. degree from Yale Law School in 1956 and is presently a partner in the law firm, Hunton & Williams, Washington, D.C., and Richmond, Virginia.

GLENN P. GIBIAN specializes in energy and environmental matters at the Kentucky Center for Energy Research, Lexington, Kentucky. He has been active in the acid rain issue for five years and in other environmental issues for seven years, he has authored several papers on environmental policy and acid rain. He is a registered professional engineer, a member of the American Pollution Control Association, and the National Society of Professional Engineers.

RANDAL H. IHARA is a Professional Staff Member of the U.S. Senate. He received his B.A. from Guilford College (1966), and his M.A. and Ph.D from the University of Tennessee (1970, 1975). Prior to accepting his present position, he served as an Instructor of Political Science at Transylvania University in Lexington, Kentucky, and then as Executive Director of the Office of Policy and Evaluation, Kentucky Energy Cabinet. He is the author of several publications on energy policy and the philosophy of social science.

TIMOTHY JOHNSON is Research Coordinator of the University of Kentucky Survey Research Center and a Doctoral Candidate in Sociology at the University of Kentucky. His research interests include social epidemiology, evaluation research, and public opinion polling.

LARRY B. PARKER is an Analyst in Energy Policy with the Congressional Research Service of the Library of Congress. He earned a B.S. and an M.S. from the Massachusetts Institute of Technology and a Ph.D. from the University of Oklahoma. His congressional research focuses primarily on utilities, coal, and environmental issues related to utilities and coal. He is the author or co-author of over three dozen reports and articles on acid rain, many of which explore economic impacts and emerging control technologies relating to the acid rain problem.

JAMES L. REGENS is an Associate Professor of Political Science and Research Fellow in the Institute of Natural Resources, University of Georgia. From 1980 to 1983, he served with the U.S. Environmental Protection Agency as Assistant Director for Science Policy, Office of Internatonal Activities (1982-83), Senior Technical Advisor to the Deputy Administrator (1981-82), and Senior Policy Analyst, Office of Research and Development (1980-81). Dr. Regens was Joint Chairman of the Interagency Task Force on Acid Precipitation (1981-82) and Chairman of the Group on Energy and Environment, Organization for Economic Cooperation and Development (1981-83). He specializes in policy analysis, regulatory impact assessment, and strategic planning.

PHILLIP W. ROEDER is an Associate Professor of Political Science and Director of the Martin Center for Public Administation and the Survey Research Center at the University of Kentucky. In addition to survey research, his research interests are in public policy analysis and state and local government.

ROBERT W. RYCROFT is an Associate Professor of Public Affairs and Political Science and Deputy Director of the Graduate Program in Science, Technology, and Public

Policy at The George Washington University. Dr. Rycroft specializes in science and technology, energy and environmental policy, and technology assessment and policy analysis. He is coauthor of: **U.S. Energy Policy: Crisis and Complacency, Our Energy Future**, and **Energy and the Western United States**, as well as articles in **World Politics, The Journal of Politics, Public Administration Review**, and **Policy Studies Review**.

DAVID G. STREETS is Manager of the Policy Sciences Section of the Energy and Environmental Systems Division at Argonne National Laboratory. He received his B.S. and PhD. degrees in physics from the University of London, England. He was a postdoctoral fellow under the sponsorship of the National Science Foundation in 1971-72 and Imperial Chemical Industries in 1972-74. He joined Argonne in 1975 and is presently responsible for management of an interdisciplinary group analyzing energy and environmental issues for the U.S. Department of Energy.

DAVID J. WEBBER is an Assistant Professor of Political Science at West Virginia University, where he teaches in its policy analysis graduate program. He has published several articles in leading policy journals, including **Policy Studies Journal** and the **Natural Resources Journal**. His current research focuses on environmental federalism.

ELIZABETH W. YANARELLA is a Masters Candidate in the Patterson School of Diplomacy and International Commerce at the University of Kentucky.

ERNEST J. YANARELLA is an Associate Professor of Political Science at the University of Kentucky. He earned his B.A. from Syracuse University in 1966 and his Ph.D. from the University of North Carolina at Chapel Hill in 1971. He is the author of over two dozen articles and and two earlier books, **The Missile Defense Controversy** (University Press of Kentucky, 1977) and **Energy and the Social Sciences** (Westview Press, 1982). His primary teaching and research interests lie in critical policy analysis (energy and environment, national security and arms control, U.S and Soviet foreign policies).

Index

Acid deposition
 aquatic damage (lakes, streams) caused by, 1-2, 5, 13-17, 88, 94, 221, 224
 aquatic biological damage (fish) caused by, 17, 88, 222
 and bureaucratic politics, 48-50, 100
 forest damage from, 1-2, 5, 18-20, 88, 224, 327; estimates of, 5; hypotheses to explain, 19. See also Federal Republic of Germany, forest damage in
 human health damage attributed to, 5, 224
 hypotheses concerning damage from, 15-17
 international diplomacy of, 261, 319, 329, 331. See also Transboundary air pollution, U.S.-Canadian impasse over
 level of public awareness of, 63-64, 87, 262, 268
 policy stalemate in U.S. over, 2, 35-36, 39-52, 100, 275, 277-286, 286-293. 318; political issues in, 287-289, 307 (n47). See also, Reagan administration, policy of postponement on acid rain
 precursors of, 4, 92; sources of, 176, 177, 223. See also Nitrogen oxides; Sulfur dioxide
 public concern over, 64-67; compared to other environmental hazards, 65-66; compared to other social issues, 6-67
 public knowledge and, 36-37, 67-70, 262; compared to other environmental threats, 36-37, 268
 public opinion on, 36-37, 46, 57-77; general findings on, 77-79. See also Regulatory strategies, public attitudes toward
 regional differences in public perception of, 70-72. See also Regional conflict over acid rain
 regions sensitive to, 14, 91, 92, 174, 175, 190, 191, 267, 290
 scientific issues concerning, 286-287, 320. See also Scientific research on acid rain in United States
 scientific knowledge of, 6-9, 81, 90, 322-323
 search for policy consensus on, 269-273, 289-298, 329; issues in achieving, 297-298
Acid precipitation
 pH of, 9, 92, 161-162; compared to natural pH, 101 (n1), 221
 See also Acid deposition
Acid rain. See Acid deposition; Acid precipitation
Adirondacks, 4, 13, 14, 17, 48, 63, 71, 137, 143, 163,

337

165, 192, 193 (figure), 195, 198, 244, 290
Alkalinity, 14, 16, 17, 28 (n38), 190. See also Acid deposition, aquatic damage caused by; Buffering capacity
American Public Power Association (APPA), 286, 295, 306 (n43)
Appalachia, 4, 5, 6, 18, 175, 177, 290
Argonne National Laboratory, 83, 165, 178, 187, 192, 198

Bagge, Carl, 114
Bilateral Research Consultation Group, 40, 88, 266. See also Transboundary air pollution, as Canadian-U.S. issue
Buffering capacity, 16, 190, 262, 301. See also Alkalinity
Burford, Anne McGill, 45
Byrd, Robert C. (Sen.), 24 (n19)

Cabinet Council on Natural Resources and the Environment, 48, 50
Camel's Hump (Vt.), 18
Canada, 40, 46, 88, 89, 147, 148, 175, 262, 264, 272, 273, 284, 299
  air pollution problems in, 75
  domestic acid rain control programs of, 239, 306-307 (n46)
  public opinion on acid rain in, 63, 65, 70, 71, 72, 73, 75-76
  transboundary proposal of, 89, 269
  vulnerability to U.S. transboundary air pollution of, 266-268
  See also Bilateral Research Consultation Group; Transboundary air pollution, as Canadian-U.S. issue
Christian Democratic Party (CDU), 250
Clean Air Act
  and amendments, 35, 107, 110-111, 114, 224-226, 275, 312 (n79), 317

"best available control technology" (BACT) provisions of, 278, 298, 311 (n74)
  issue of total loadings vs. maximum concentration of pollutants in, 111
  reauthorization efforts for, 1, 3-4, 11, 41, 87, 100, 107
  See also National Ambient Air Quality Standards; New Source Performance Standards, State Implementation Plan
Coal cleaning (pre-combustion), See Technology, emission control
Coal industry
  high-sulfur vs. low-sulfur facets of, 46, 286, 287, 293, 299
  impact of control programs on markets of, 83, 189-190, 241, 272, 277-278
Commoner, Barry, 46
Computer modeling, 83, 177
  AIRCOST, 178, 187, 192
  ASTRAP, 192, 329
  AUSM, 178
  CEUM, 178
  IBM, 183
  ICE, 183
  long-range transport, 138, 198
  MCARLO, 148, 149, 161
  MEP, 163
  MINCOST, 181, 187
Conservation Foundation, 219
Cost-benefits analysis, 2, 94, 162-164, 320-321, 322
  as ideology, 47, 50-51
  and political choice, 100-101, 320-321
Cowling, Ellis, 115

Department of Energy, 48, 100, 282

Earth
  as an ecosystem, 252-253
  strategic vulnerabilities of, 253
Ecology, 36, 240, 243-244
  and acid rain problem, 252-255
  See also Green party
Economic Commission for Europe (ECE), 88, 89, 263
  "30 % Club" established

among members of, 89, 101, 239, 241, 270, 271, 301
Emission reduction strategies
Aspin bill, 211-215, 229
bubble policies, 205-211, 231, 329, 330
Coalition of Northeastern Governors proposal, 285
D'Amours bill, 146, 161, 165, 185, 189, 281
Durenberger bill, 229, 279, 295, 306 (n43)
National Governors Association proposal, 285, 295, 305 (n7)
New England Caucus bill. See D'Amours bill
Rinaldo bill, 229
Senate Committee bill, 4, 21 (n4), 89, 145, 157, 160, 165, 277, 279
state-based economic-incentive policy, 232-236
Udall-Cheney bill, 185
uniform percentage reduction strategy, 146-147, 160, 294
Waxman-Sikorski bill, 4, 21 (n4), 21 (n5), 89, 146, 148, 161, 165, 185, 189, 196, 227, 229, 281
See also Regulatory strategy
Emissions growth CAP, 277, 285, 298
Energy Supply and Environmental Coordination Act of 1974, 113-114
Environmental Defense Fund, 246
Environmentalism, 36, 52, 240, 243-244
professionalization of, 44, 245
right-wing critique of, 243
Environmental movement
ecological lessons for, 255
limitations of, 36, 39-40, 240, 257 (n17)
as middle-class movement of moral protest, 245
preservationism as forerunner of, 245
progressive conservationism as precursor of, 244-245
Environmental protection
bureaucratization of, 36, 50
Environmental Protection Agency, 36, 40, 45, 100, 207, 231, 254, 277, 282
interim control policy proposed by, 47-48, 50
possible role in future acid rain control policy by, 96, 279, 280, 300, 301
See also Environmental protection, bureaucratization of; Ruckelshaus, William
Environmental threats
progress in reducing, 219, 220
shift in public attitudes toward, 46
See also Acid deposition, public knowledge of; Acid deposition, public concern over
European Community, 264, 269

Federal courts, 300-301
Federal Emissions Options Market, 233, 235, 236
Federal Republic of Germany, 88, 89, 101 (n2), 247, 248, 249, 250, 261, 262, 264
acid rain policy of, 250, 251, 262
attachment to nature evidenced in, 247
forest damage in, 18, 239, 250, 327
flue gas desulfurization technology (FGD). See Technology, emission control
Friends of the Earth, 246
Fuel-switching, 5, 92, 96, 180, 189, 213, 214, 241, 279. See also Coal industry
Funding strategies, 95-98, 297
cost-sharing issue in, 22 (n5), 95, 98, 285, 292-293, 295, 296 (figure), 297
emissions tax, 234, 235, 285
fuels tax, 297
role of utility rate-making process in, 91, 96, 98
trust fund, 22 (n5), 48, 82, 297, 328

Glenn, John (Sen.), 95, 279

Great Britain, 261, 262
Green party, 240, 244, 255
　acid rain proposals of, 250
　compared with American environmental movement, 251
　political platform of, 249
　search for new ecological paradigm by, 249
　See also Ecology; Federal Republic of Germany

Hodel, Donald, 282. See also Department of Energy

ICF, Inc., 161, 183, 189, 195
Interest-group liberalism (or politics), 36, 43, 51, 88, 100, 246
International Nickel Company (INCO), 75, 76
Izaak Walton League, 244

Love Canal (N.Y.), 47, 68, 70
Lovelock, James, 252
Lugar, Richard (Sen.), 7, 292, 293

Merchant, Carolyn, 254
Mitchell, George (Sen.), 7, 279
Mohnen, Volker, 26 (n28), 45
Mt. St. Helens, 69
Mumford, Lewis, 254, 255

National Acid Precipitation Act (1980), 1, 40, 112, 175
National Acid Precipitation Assessment Program (NAPAP), 161, 178, 183, 284, 290, 292
National Audubon Society, 245
National Clean Air Coalition, 243
National Academy of Sciences report (1983), 2, 46
　findings of, 12, 26 (n33), 77, 78, 95, 290, 291, 292, 293, 309 (n12)
National Ambient Air Quality Standards (NAAQS), 84, 109-111, 113, 119, 204, 225. See also Clean Air Act

National Research Council report (1983). See National Academy of Sciences report (1983)
National Wildife Federation, 219, 243
Natural Resources Defense Council, 111, 243, 246
New Source Performance Standards (NSPS), 11, 84, 113, 136 (n13), 185, 200, 205, 282
　ability to control acid rain threat through, 113-116, 117, 119, 124-135, 204
　See also Clean Air Act
Nitrogen oxides (NO_x)
　emissions of, 10, 11, 92, 176
　sources of, 11
　estimates of future emissions of, 224
Nixon administration, 36, 49
Nordic Convention on the Protection of the Environment, 265
Nuclear power, 115, 126, 129, 131

Office of Management and Budget (OMB), 48, 50. See also Stockman, David
Office of Science and Technology Policy (OSTP)
　interim report of, 46, 252, 253
Office of Technology Assessment, 223
Oppenheimer, Michael, 45
Organization for Economic Cooperation and Development (OECD), 88, 264-265, 269
Ozone, 19, 254

Politics of growth, 44
Public opinion
　and doorstep opinions, 58
　and environmental issues, 57
　general findings on, 57-59
　level of awareness expressed in, 59-60
　level of knowledge indicated by, 60
　role in policymaking of, 57
　See also Acid deposition,

341

public opinion on;
Acid deposition, public knowledge of; Acid
deposition, level of
public awareness of;
Acid deposition, public concern over

Reagan administration
  and the administrative
    presidency, 49-50
  policy on environmental
    protection, 48
  policy of postponement on
    acid rain control, 51,
    76, 282-284, 298, 299
  See also, Transboundary
    air pollution, as
    Canadian-U.S. issue,
    United States, anti-control alliance in;
    United States, politics of acid rain in
Regional conflict over acid
  rain, 3, 4, 36, 44,
  50, 71, 90, 175, 285
  286, 287, 307 (n48),
  288 (figure). See also
  Risk assessment, politics of
Regulatory strategies
  applicability of "polluter
    pays" principle in, 82,
    96, 310 (n68)
  economic costs of, 81, 82,
    94, 162, 163
  emissions targets approach as one type of,
    96, 137
  employment impact of, 6,
    50
  environmental benefits
    of, 81, 82, 95, 137,
    162, 163
  equity vs. efficiency
    issue in, 82, 83, 99-
    100, 163, 164, 165
  goals of, 289-292
  impact on electric utility rates of, 6, 23
    (n12), 23 (n13), 94, 96,
    97
  least-cost approach of,
    162, 163, 180, 181, 182,
    186 (figure)
  public attitudes toward,
    72-77
  role of market-based programs in, 83, 102 (n7),
    220-221, 229-232
Risk assessment

politics of, 1, 4-9
See also Regional conflict over acid rain
Ruckelshaus, William, 39,
  47, 78, 282, 283, 284,
  297, 304 (n26), 305 (n32)

Scientific research on acid
  rain
  in United States, 9, 15,
    27 (n37), 173, 321, 325,
    326, 330, 331; shortcomings of, 9, 253, 309
  in Europe, 18, 32 (n56),
    32 (n58), 32 (n60), 253,
    262
  role in policy impasse,
    45-46
  See also Acid deposition,
    scientific issues concerning; Acid deposition, scientific
    knowledge in
Scrubbers. See Technology,
  emission control
Sierra Club, 245
Smith, Robert Angus, 40, 90
Social Democratic Party
  (SPD), 248, 250
Source-receptor issue,
  2, 11-13, 25 (n28), 175
  linearity vs. non-linearity issue involved in, 266 (n30),
    287, 291, 326
  impact on control strategies of, 190, 287
Stafford, Robert (Sen.), 4,
  7, 278, 307 (n48). See
  also Emission reduction
  strategies, Stafford bill
State Implementation Plan
  (SIP), 109, 110, 225,
  235, 236, 277, 319.
  See also Clean Air Act
Stockman, David
  views on acid rain, 47
  opposition to interim EPA
    proposal, 49, 282
  See also Office of Management and Budget
Sulfur dioxide (SO_2)
  emissions of, 9, 10, 11,
    92, 176
  estimates of future emissions of, 223
  top 50 emitters of, 101
    (n5), 226 (table), 227
Sweden, 88, 101 (n2), 261,
  263, 265
  acid rain research on,

88, 262
as leader in tackling acid rain problem, 262, 264
lake liming strategy of, 275, 301
See also Scientific research on acid rain, in Europe

Tall smokestacks policy, 41, 112, 225
Technology, emission control
coal cleaning (pre-combustion), 92, 96, 180, 241, 272
flue gas desulfurization (FGD), 90, 92, 96, 180, 181, 184, 196, 213, 214
"scrubbers," 5, 21, 96, 121, 189, 213, 271, 279, 281, 282, 294, 295, 298, 304 (n18), 329
Tennessee Valley Authority (TVA), 286
Three Mile Island, 46, 70
Times Beach (Missouri), 46
Transboundary air pollution
as Canadian-U.S. issue, 1, 40, 239, 241, 266-269, 270, 284, 299, 318, 321-322, 323, 324, 325, 327
as European issue, 239, 261, 262-265
See also Canada, transboundary proposal of; Canada, vulnerability to U.S. air pollution of
Transfer matrices, 138, 139, 192, 195
Trisko, Eugene, 112, 192

United Mine Workers, 50, 287
United Nations Conference on the Human Environment (1972), 85
declaration of, 219
United States
anti-control alliance in, 41, 42 (figure), 270, 275; strategy of, 41; shift in strategy by 51; weakening of, 46-47. See also Acid deposition and bureaucratic politics; Reagan administration
politics of acid rain in, 39-53, 277-302
pro-control alliance in, 41, 42 (figure), 43, 46, 270, 275; strategy of, 41, 42, 43; shortcomings of 52-53. See also Environmental movement
two-party system, role in policy impasse, 44-45; narrow ideological spectrum of, 253; compared to multi-party system, 251, 253
U.S. Congress
acid rain politics in, 3-4, 89-90, 173-177, 183-190, 277-282, 289, 307 (n47)
See also Emission reduction strategies
U.S. Geological Survey (U.S.G.S.), 14-15
Utility industry
views on acid rain of, 286
impact of varying rates of retiring coal-fired plants by, 117, 121, 124-135, 200-204, 302, 313 (n84)
See also Sulfur dioxide, emissions of

Waxman, Henry (Rep.), 4, 281. See also Emission reduction strategies, Waxman-Sikorski bill
West Germany. See Federal Republic of Germany
Wolfe, Alan, 48